Characterization of Biological Systems *via* Relaxometric and Diffusimetric NMR

A dissertation submitted to the faculty of
WORCESTER POLYTECHNIC INSTITUTE
in partial fulfillment of the requirements for the degree of
Doctor of Philosophy
in
Biomedical Engineering

May, 1998

by

Sam S. Han

Approved:

Christopher H. Sotak, Ph.D.
Professor, Major Advisor
Biomedical Engineering Department
Worcester Polytechnic Institute

Karl G. Helmer, Ph.D.
Research Assistant Professor
Biomedical Engineering Department
Worcester Polytechnic Institute

Katsumi Irie, M.D.
Neurologist
Department of Neurology
Memorial Hospital,
The University of Massachusetts Medical School

Peter Grigg, Ph.D.
Professor
Department of Physiology
The University of Massachusetts Medical School

Robert A. Peura, Ph.D.
Professor and Head
Biomedical Engineering Department
Worcester Polytechnic Institute

*The beginning of wisdom
is the fear of the LORD,
and knowledge of the Holy One
is understanding.*

Proverbs 9:10 (NIV)

Acknowledgements

As I look back on my academic career, there are many people who have spurred me on to give my best. It is these people to whom I owe an outstanding debt of gratitude.

First and foremost, I have to give all glory, honor, and praise to my God and Savior, Jesus Christ, from whom all blessings flow. Indeed, my strength and inspiration comes from God alone. Also, I thank God for bringing people into my life who have been a constant source of encouragement. Of these people, I am most grateful to my parents who have never stopped loving me from my diaper days to my dissertation days. Their support is something that I will never be able to repay. Amy Wong (soon to be Mrs. Sam Han), who has been by my side through thick and thin for the past six years, has been a true woman of noble character. In her support and encouragement, she ranks right up there with my parents.

As my advisor, mentor, and colleague, Professor Christopher Sotak has been an integral part of shaping my life as a scientist. In the area of academics, honor, and diligence, I have met no equal. Never once has he expected anything less than my best efforts. Truly, he is a man of integrity and I have learned much from his wellspring of wisdom. Dr. Robert Peura, my department head, has been a true example to me in the areas of diplomacy. I covet his skills and have many times been the beneficiary of these skills. Dr. Peter Grigg has not only taught me some of the fundamental thought processes

ACKNOWLEDGEMENTS

behind asking scientific questions but has also demonstrated that it is possible to maintain my humanity in the midst of working like a machine to collect and process data. For this I thank him.

I would like to thank my predecessor, Dr. Bernard Dardzinski, for allowing me to work alongside him in several of his projects, and for instilling in me good work habits. Dr. Lawrence Latour, who has taught me to never lose sight of the big picture, has been integral in shaping my thought processes and I have profited from his intuitive grasp on difficult concepts. I thank Dr. Karl Helmer for his willingness to help in all situations and for providing avenues to further my knowledge even at the "bonus baby" stage of my graduate studies. I will always remember Dr. Katsumi Irie as a good doctor, a good scientist, and a good friend. I thank Matt Silva for bearing patiently with me in my greatest times of victory and my worst times of defeat in our joint ventures. Stephen Gemmell, Rick Carano, Mike Meiler, Fuhai Li, Kentaro Takano, Turgut Tatlisumak, and Cy Tamanaha, I thank for encouraging me during various stages of my life at Worcester Polytechnic Institute.

Finally, I thank my extended family at church that has kept me accountable in all other areas of my life.

Preface

The following manuscript is the product of my efforts to understand a fraction of the mind of God.

From the beginning, I had the great fortune of adopting one of my predecessor's final projects as my first project. Working on this project became a milestone in my studies, not only as an initial project, but more so because of the experience of working with an adept NMR scientist. The preliminary hypothesis of this study was to observe changes in murine tumor oxygenation with radiotherapy. In preceding works, it was shown that the spin-lattice relaxation rate of perfluorocarbons was directly proportional to the dissolved oxygen tension. This relationship between fluorine relaxation and oxygenation, along with NMR imaging techniques, allowed for evaluation of therapeutic interventions using perfluorocarbons.

During this time, a parallel experiment was in progress to observe the effects of oxygenation on the diffusion of water molecules in murine tumors. Other groups had shown that there was a highly positive correlation between tumor oxygenation and the apparent diffusion coefficient (ADC) of water in the tumor. These findings were troubling because intuitively, if there were any correlation between tumor oxygenation and ADC, we would expect to see a negative rather than a positive correlation. To

PREFACE

resolve this issue, we undertook the project of experimentally correlating tumor oxygenation using ^{19}F imaging and ^{1}H ADC mapping techniques. This proved to be a rewarding experiment because our intuition served us correctly. Despite the findings by other groups, our studies showed that no apparent correlation existed between ADC values and tumor oxygenation.

Studies in animals brought up questions about the behavior of water in the brain following the onset of stroke. During cerebral ischemia, ionic imbalances in the brain result in changes in cellular water distribution and give rise to edema (cellular swelling). Although it is clear that the properties of diffusing water are correlated with the swelling of cells, it was difficult to understand the intricacies behind the relationship of water to its surroundings. The answer to this basic question of what is happening to the water in a complicated two-compartment system has been one of the motivating factors for my study of water in a model system of yeast cells. Studies of two-compartment systems by others have shown non-monoexponential diffusion signal attenuation, which has been assumed to be because of the different compartmental contributions. Although this may be a possibility, previous works have shown that even in single compartment systems, non-monoexponential behavior could be observed simply as a result of the tortuous diffusion pathways of water in the interstitial space. This observation in single compartment systems cast doubt on the interpretation of the data from the two-compartment system. Not only are there compartmental effects on diffusion, but the diffusion behavior is also confounded by restriction effects. One of the premises behind

PREFACE

the work done in yeast-cell suspensions was that a deconvolution of the compartmental effects could provide valuable information on the behavior of water. With this goal in mind, a majority of my research efforts was devoted to investigating model systems for understanding the diffusion behavior of water in a two-compartment system. Fortunately, a well-characterized system was available in yeast-cell suspensions. The yeast experiments should serve merely as a stepping stone to other experiments which further characterize the model system and, ultimately, carry over to *in vivo* experiments to understand the behavior of water during ischemic brain injury.

The last experiment in this dissertation deals with diffusion of water in excised rabbit Achilles tendon. The initial set of experiments on tendon gave very promising results and eventually led to a Master's thesis for one of my colleagues. Unfortunately, further investigation into the behavior of the tendon revealed that these observed effects were no more than an artifact of the storage medium. The initial discovery of this fact was disappointing, but at the same time opened up another avenue for the direction of these experiments. In wrestling with the logistics of measuring NMR properties of tendons, most of the groundwork has been laid for future studies. Although these results are not completely revealing, we are still of the belief that the viscoelastic properties of tendons change with tensile load. Whether this phenomenon is observable *via* NMR is still to be determined.

PREFACE

Although diverse in nature, all of these experiments have taught one common lesson. Ask the good question and seek the right answer. I hope that the writings herein will be a reflection of the work done to seek the right answer.

May God bless the reading of this dissertation.

Abstract

NMR techniques have seen widespread use in the characterization of biological systems. Due to the different environments experienced by spins, the T_1 and T_2 relaxation behavior can differ significantly. Furthermore, the local environment can be altered by the introduction of paramagnetic agents that alter the bulk magnetic susceptibility of a sample, causing drastic changes in the relaxation behavior of spins. For spins situated in restricted spaces, the diffusion characteristics as well as the relaxation properties of these spins can differ from system to system.

Normally, the relaxation (T_1 and T_2) and diffusion measurements are performed independently and system parameters are determined solely on the basis of the independent behavior of the relaxation or diffusion. In the following experiments, these parameters (T_1, T_2, and apparent diffusion coefficient) were used to characterize biological systems. Furthermore, multi-dimensional analysis was performed on a system where combined relaxometry and diffusimetry were used to characterize the system as a function of both parameters.

Radiotherapy studies in mice using perfluoro-15-crown-5-ether showed a decrease in pO_2 following a single large dose of radiation. In conjunction with spectroscopic data, Inversion-Recovery Echo-Planar-Imaging data were collected at 1-3 hours, 10-13 hours, and 19-26 hours post irradiation, and T_1-maps generated in order to display localized

ABSTRACT

changes in pO_2. The calculated T_1-maps were then weighted by their respective M_0-maps to find the weighted average of the T_1-maps, and an equivalent pO_2 of the tumor was then calculated from the weighted average. Untreated control animals that were subjected to the same time course showed no evidence of pO_2 decline, while the tumors irradiated with a single dose of 6 MeV electrons showed a decline in pO_2 by approximately 9 torr almost immediately after irradiation. The calculation of pO_2 using the weighted average of the T_1-maps was not only highly correlated to the spectroscopic measurements, it was approximately equivalent to the spectroscopic measurements. It is speculated that the decrease in the tissue oxygenation following radiation therapy is due to vascular damage caused by such a high dose of radiation, or edema within the interstitium of the tumor. Edema can cause the interstitial pressure to increase, resulting in vascular collapse. This in turn would lead to decreased perfusion and thus decreased oxygen delivery.

Water diffusion-coefficient mapping was used in conjunction with ^{19}F inversion-recovery echo-planar imaging (IR-EPI) of a sequestered perfluorocarbon (PFC) emulsion to investigate the spatial correlation between the diffusion coefficient of water and the tissue oxygen tension (pO_2) in radiation-induced fibrosarcoma (RIF-1) tumors (n = 11). The diffusion-time-dependent apparent diffusion coefficient, $D(t)$, was determined by acquiring diffusion coefficient maps at 20 different diffusion times. Maps at four representative time points in different regions of the $D(t)$ curve were selected for final analysis. An intravenously administered PFC emulsion, perfluoro-15-crown-5-ether, was used to generate the pO_2 maps. $D(t)$ and pO_2 data were acquired with the animal

ABSTRACT

breathing either air or carbogen (95% O_2 – 5% CO_2) to investigate the effects of increased tumor pO_2 on $D(t)$. The average increase in tumor pO_2 was 22 torr when the breathing gas was changed from air to carbogen. Correlating plots generated from pixel data for $D(t)$(air breathing) versus $D(t)$(carbogen breathing) showed little deviation from a slope of unity. Correlation plots of $D(t)$ versus pO_2 indicate that no correlation is present between these two parameters. This study also confirms that necrotic tissue was best differentiated from viable tumor tissue based on $D(t)$ maps at long diffusion times.

Diffusion-signal-attenuation curves in yeast-cell suspensions show non-monoexponential signal decay that is assumed to arise from separate compartmental contributions to the overall signal. However, restricted diffusion effects also give rise to non-monoexponential signal decay and are difficult to separate from compartmental signal contributions. Combined relaxometry and diffusion measurements allows differentiation between compartmental diffusion constants by first separating the compartmental contributions on the basis of differences in their respective relaxation times. Diffusion-weighted inversion-recovery spin-echo experiments were carried out at different b-values. Intra- and extracellular compartments in yeast-cell suspensions were separated on the basis of T_1 relaxation by adding an MR contrast agent to the extracellular space. Once the compartmental signals were distinguished on the basis of T_1, the relative signal attenuation for each compartment was used to calculate the separate compartmental ADCs. With this method, even compartmental diffusion coefficients with similar values can be distinguished.

ABSTRACT

Water diffusion measurements were made on rabbit Achilles tendon to determine their behavior during static tensile loading and unloading. Tendons previously stored frozen in phosphate-buffered saline (PBS) were sequentially loaded with 0.4, 5, 10, and 0.4 N loads. The apparent diffusion coefficient (ADC) was measured perpendicular ($ADC_\perp$) and parallel ($ADC_\parallel$) to the fiber orientation at diffusion times of 10, 30, and 60 ms for each load. $ADC_\perp$ and $ADC_\parallel$ increased with increasing load for all samples. New samples of freshly-harvested tendons were loaded with a static 5-N load for five minutes and then unloaded for 30 minutes to determine the effects of loading and unloading. The $ADC_\perp$ of fresh tendon was studied as a function of loading and unloading. The $ADC_\perp$ increased with load for all samples. This increase is attributed to the extrusion of tendon water into a bulk phase outside the tendon. The recovery of the tendon upon unloading exhibited a reversal of the $ADC_\perp$ back to the baseline value. This recovery is attributed to the water moving from the bulk phase to the bound phase. The recovery followed a slower time course than the extrusion of water. It was also found that the phosphate-buffered saline caused the tendon to swell. This method can be used both to detect structural changes in tendon under tensile loading and to study the transport of water in tendon.

Table of Contents

Acknowledgments i

Preface iii

Abstract vii

Table of Contents xi

List of Figures xiv

List of Tables xvi

1. **NMR Theory** 1
    1.1. The Electromagnetic Field 1
    1.2. Nuclear Spin 2
    1.3. Rotating Frame of Reference and Radio-Frequency Pulses 8
    1.4. The Free Induction Decay (FID) 11
    1.5. Fourier Transform and the Frequency Domain 12
    1.6. Relaxation Time Constants 17
        1.6.1. T_1 Relaxation (Spin-Lattice or Longitudinal Relaxation) 17
        1.6.2. T_2 Relaxation (Spin-Spin Relaxation) 21

2. **Spin Manipulation** 25
    2.1. The Chemical Shift and Binomial Pulses 25
    2.2. Spin Echoes (SE), Stimulated Echoes (STE) and Spurious Echoes ... 28
    2.3. Magnetic Resonance Imaging 31
        2.3.1. Effect of Magnetic Field Gradients on Spins 31
            2.3.1.1. Frequency Encoding 32
            2.3.1.2. Slice Selection 33
            2.3.1.3. Phase Encoding 36
        2.3.2. k-space and Imaging 37
            2.3.2.1. Two-Dimensional Fourier Imaging 38
            2.3.2.2. Saw-Tooth Echo-Planar Imaging 40
    2.4. Diffusion 42
        2.4.1. The Diffusion Process and Stochastic Modeling 42

    2.4.2. Measuring Diffusion with Pulsed-Field Gradient NMR 44

3. Murine Tumor ^{19}F Experiments **50**
  3.1. Radiation Therapy 51
    3.1.1. Abstract 51
    3.1.2. Introduction 52
      3.1.2.1. Oxygen Sensitive ^{19}F MRI and MRS 53
      3.1.2.2. Hypoxia and Radioresistance 54
    3.1.3. Experimental Methods 56
      3.1.3.1. Hardware and Materials 56
      3.1.3.2. Animal Preparation 56
      3.1.3.3. Data Acquisition 57
      3.1.3.4. Radiation Therapy 58
      3.1.3.5. Data Analysis 59
    3.1.4. Results 60
    3.1.5. Discussion and Conclusion 65
    3.1.6. Acknowledgements 68

4. Murine Tumor Diffusion and pO_2 Experiments **69**
  4.1. On the Correlation Between the Water Diffusion Coefficient and Oxygen Tension in RIF-1 Tumors 70
    4.1.1. Abstract 71
    4.1.2. Introduction 72
    4.1.3. Background 75
    4.1.4. Methods 76
    4.1.5. Results 81
    4.1.6. Discussion 88
    4.1.7. Acknowledgements 96

5. Yeast Experiments **98**
  5.1. Deconvolution of Restriction Effects on Compartmental Diffusion Using Combined Relaxometric and Diffusimetric NMR 98
    5.1.1. Abstract 98
    5.1.2. Introduction 99
    5.1.3. Theory 102
    5.1.4. Methods 108
      5.1.4.1. Yeast Preparation 108
      5.1.4.2. NMR Experiments 109
      5.1.4.3. Data Analysis 111
    5.1.5. Results 113
      5.1.5.1. [Gd-DTPA] and Relaxographic Peak Separation 113

		5.1.5.2. Inversion-Recovery Experiments with Varying Gradient Strengths and Echo Times 114

                      5.1.5.3. Calculation of the Biexponential Diffusion Decay Constants from the PFGSE Data 117

                      5.1.5.4. Calculation of the ADCs Associated with the Fast and Slow Relaxing Components of the Signal 117

        5.1.6. Discussion 121

6. Rabbit Achilles Tendon Experiments 125

    6.1. Characterization of Water ADC Behavior Under Tensile Loading and Recovery in Rabbit Achilles Tendon Using NMR 126

        6.1.1. Abstract 127

        6.1.2. Introduction 128

        6.1.3. Methods 129

                6.1.3.1. Equipment and Apparatus 129

                6.1.3.2. Load Dependence of Tendon Water ADC 130

                6.1.3.3. Diffusion-Time Dependence of the ADC 131

                6.1.3.4. Fast Time Measurements of Water ADC in Response to Uniaxial Tensile Loading 132

                6.1.3.5. Studies of Tendon Mechanical Behavior 134

        6.1.4. Results 135

                6.1.4.1. Load Dependence of Tendon Water ADC 135

                6.1.4.2. Diffusion-Time Dependence of the ADC 136

                6.1.4.3. Fast Time Measurements of Water ADC with Loading and Unloading 139

                6.1.4.4. Tendon Anisotropy and Effects of Storage Media 141

                6.1.4.5. Mechanical Response of Tendon to Loading 142

        6.1.5. Discussion 143

7. Summary 150

8. Bibliography 154

9. Appendix A.1

List of Figures

Chapter 1
1.2.1.	The angular momentum of a spin-$\frac{1}{2}$ nucleus	3
1.2.2.	Spin-$\frac{1}{2}$ nuclei in a static magnetic field, $\vec{B}_0$	4
1.2.3.	The net magnetization	8
1.4.1.	The free-induction decay (FID)	11
1.5.1.	Fourier pairs commonly used in NMR	15
1.5.2.	The FID and the Fourier transform (FT) of the FID	16
1.5.3.	Sinc and rectangular wave Fourier pair	16
1.6.1.1.	Inversion of spins upon application of a 180° pulse	18
1.6.1.2.	Inversion-recovery (IR) pulse sequence	19
1.6.1.3.	Inversion-recovery (IR) curve	20
1.6.2.1.	Hahn spin echo (SE) pulse sequence	22
1.6.2.2.	Spin behavior during Hahn spin echo pulse sequence	23

Chapter 2
2.1.1.	1-3-3-1 binomial pulse	26
2.1.2.	Illustration of a 1-3-3-1 chemical-shift-selective binomial pulse sequence on the net magnetization of water and fat	27
2.2.1.	Stimulated-echo (STE) pulse sequence	30
2.3.1.1.	Relative field strength as a function of spatial position	31
2.3.3.1.	Sinc and rectangular wave Fourier pair	35
2.3.4.1.	Phase-encoding gradient	36
2.3.5.1.	Conventional two-dimensional Fourier imaging k-space raster	38
2.3.5.2.	Gradient-recalled echo (GRE) pulse sequence	39
2.3.5.3.	Spin-echo echo-planar imaging (SE-EPI) pulse sequence	41
2.3.5.4.	Saw tooth k-space trajectory for echo-planar imaging	41
2.4.2.1.	Stejskal-Tanner pulsed-field gradient spin-echo (PFGSE) pulse sequence	48

Chapter 3
3.1.	Plot of spectroscopically determined ΔpO_2 as a function of time post-irradiation with a single large dose (1000 cGy) of 6 MeV electrons	62
3.2.	Plot of ΔpO_2 determined using the weighted-average of the T_1-maps as a function of time post irradiation with a single large dose (1000 cGy) of 6 MeV electrons)	63
3.3.	pO_2 maps of typical tumors during the time course of the radiation therapy study	64

Chapter 4

4.1.	Examples of oxygen tension maps in a RIF-1 tumor as a function of breathing gas	82
4.2.	Example of a histological slide used to determine necrotic regions for RIF-1 tumors	84
4.3.	Schematic representation of a typical $D(t)$ versus $t^{½}$ curve for a RIF-1 tumor showing the four maps used in the analysis (out of the twenty acquired)	85
4.4.	Scatter plots for the shortest and longest diffusion times showing the correlation between $D(t)$ for air breathing and $D(t)$ for carbogen breathing for a single RIF-1 tumor	86
4.5.	Scatter plots for air and carbogen breathing showing the correlation between $D(t)$ and pO_2 for the longest diffusion time for a single RIF-1 tumor	87

Chapter 5

5.1.4.1.	Inversion-recovery spin-echo pulse sequence with diffusion gradients	110
5.1.4.2.	Pulsed-field-gradient spin-echo (PFGSE) pulse sequence	111
5.1.5.1.	Inversion-recovery curves at different gradient values	114
5.1.5.2.	Inversion-recovery curves at different echo times	115
5.1.5.3.	Diffusion attenuation curve for yeast-cell suspension acquired using PFGSE pulse sequence in Fig. 5.1.4.2	118
5.1.5.4.	Calculated M_{0a} and M_{0b} from water in yeast-cell suspensions plotted as a function of b-value	119
5.1.5.5.	Normalized plot of the calculated fits using the two different methods	120

Chapter 6

6.1.	Loading protocol for rabbit Achilles tendon	133
6.2.	Load dependence of $ADC_\parallel$ and $ADC_\perp$ for applied loads of 0.4, 5, 10, and 0.4 N	136
6.3.	Load dependence of $ADC_\perp$ for applied loads of 0.4, 5, 10, and 0.4 N at 10, 30, and 60 ms diffusion times	137
6.4.	The diffusion-time dependence of water $ADC_\perp$ and $ADC_\parallel$	138
6.5.	Response of a fresh tendon to static 5-N loading	139
6.6.	Response of a fresh tendon to static 5-N loading and unloading	140
6.7.	Average tendon response curve to static 5-N loading and unloading (n = 8)	140
6.8.	Comparison of diffusional anisotropy for tendons stored frozen in phosphate buffered saline and freshly harvested tendons	141
6.9.	Stress-strain curve for an Achilles tendon using loads up to 10 N	143
6.10.	Photograph showing the extrusion of bulk water from a rabbit Achilles tendon (stored in phosphate-buffered saline) under a 10-N load	144

List of Tables

Chapter 3

3.1. Spectroscopically determined change in tumor oxygen tension (ΔpO_2) from baseline values measured at the indicated time periods post-irradiation (1000 cGy of 6 MeV electrons) for the four animals given radiation therapy (Rad1, Rad2, Rad3, and Rad4) and two control animals (Cont1, Cont2) 60

3.2. Changes in tumor oxygen tension (ΔpO_2) from baseline values measured at the indicated time periods post-irradiation (1000 cGy of 6 MeV electrons) for the four animals given radiation therapy (Rad1, Rad2, Rad3, and Rad4) and two control animals (Cont1, Cont2) 61

3.3. Summary of tumor pO_2 changes (ΔpO_2) measured spectroscopically and using the weighted average of the T_1-maps 61

Chapter 4

4.1. Changes in tumor pO_2 with a change in breathing gas from air to carbogen for 11 RIF-1 tumors 83

4.2. Fitting parameters for correlation plots of $D(t)$ for air breathing versus $D(t)$ for carbogen breathing for RIF-1 tumors 84

Chapter 5

5.1.5.2.1. Calculated T_1 and M_0 values for the fast and slow components of relaxation at different b-values 116

5.1.5.2.2. Calculated T_1 and M_0 values for the fast and slow component of relaxation at different echo times 117

Chapter 6

6.1. Percent change in ADC for tendons stored frozen in phosphate buffered saline and freshly harvested tendons 142

Chapter 1

NMR Theory

In order to understand NMR, one needs to have a basic understanding of the principles behind how an NMR signal is generated and manipulated. This chapter briefly covers NMR fundamentals, from the origins of the signal, to the acquisition and processing of the signal. There are numerous texts available which give a more detailed discussion of these principles. For the purposes of this dissertation, only the relevant terms will be introduced and discussed. More intricate details of the NMR phenomena and terms relating to it can be found in texts by Callaghan (1991), Abragam (1961), and Morris (1985).

1.1. The Electromagnetic Field

In classical electrodynamics, the magnetic field, $\vec{H}$, induced by a current, I, in an n-turn circular conductor of radius, r, is given by (Weast, 1971)

$$\vec{H} = \frac{2\pi nI}{r}. \qquad [1.1.1]$$

The above equation holds for a field produced *via* electrical induction and can be modified to include the effects of the sample's bulk susceptibility, χ_m, to give the expression for the total magnetic field as (Schwarz, 1990)

$$\vec{B} = \mu_0 (1 + \chi_m) \vec{H} \qquad [1.1.2]$$

where μ_0 is the permeability constant of free space. A careful study of Eq. [1.1.2] shows that depending on the susceptibility of the sample, the field strength within the sample can be greater than the applied field. Many authors choose to call $\vec{H}$ the magnetic field and $\vec{B}$ the flux density. But in classical magneto-statics, $\vec{B}$ is indisputably the fundamental quantity so, in this work, $\vec{B}$ will be referred to as the magnetic field.

1.2. Nuclear Spin

Although the ensemble average of nuclei are observed in NMR, a basic understanding of the individual nucleus gives us insight as to what is happening as a whole.

A fundamental tenet of quantum mechanics states that the angular momentum ($\vec{P}$) of a nucleus can only take on discrete (quantized) values given by

$$\vec{P} = \hbar \sqrt{I(I+1)} \qquad [1.2.1]$$

where $\hbar$ is Planck's constant divided by 2π and I is the total nuclear spin quantum number ($I = 0, \frac{1}{2}, 1, \frac{3}{2}, \ldots$). Spin is an inherent property of all nuclei, and the total nuclear spin is the result of the pairing of the spin for each nucleon in the nucleus. In order for a nucleus to be NMR observable, it must have a non-zero value of I, which implies an incomplete spin pairing of protons and/or neutrons. In addition to the magnitude, the orientation of the angular momentum vector is quantized. This implies that the z-component of the angular momentum is limited to values given by

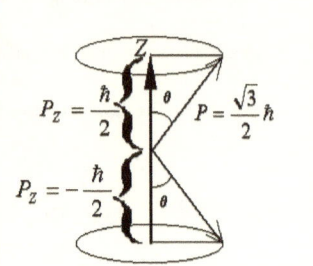

Fig. 1.2.1. The angular momentum of a spin-$\frac{1}{2}$ nucleus can have only two possible orientations.

$$P_z = \hbar m_i \qquad \text{for } m_i = -I, -I+1, \ldots, I-1, I \qquad [1.2.2]$$

where m_i is referred to as the directional (or magnetic) spin number. For a spin-$\frac{1}{2}$ nucleus, the possible orientations of $\vec{P}$ are represented in Fig. 1.2.1, where $\|\vec{P}\| = \frac{\sqrt{3}}{2}\hbar$, $P_z = \pm\frac{\hbar}{2}$, and $\theta = \cos^{-1}\left(\frac{P_z}{\|\vec{P}\|}\right)$.

The nuclear dipole moment, $\vec{\mu}$, is related to the nuclear spin angular momentum by the gyromagnetic ratio, γ, of the specific nucleus (Eq. [1.2.3]).

NMR THEORY

$$\vec{\mu} = \gamma \vec{P} \quad [1.2.3]$$

The gyromagnetic ratio is proportional to the charge-to-mass ratio of the nucleus and is, therefore, a unique value for each element. Since $\vec{P}$ is quantized (Eq. [1.2.1]), then by association through Eq. [1.2.3], $\vec{\mu}$ must also possess only discrete values.

When nuclei are placed in an external magnetic field, $\vec{B}_0$, the potential energy, E, of the interaction between $\vec{\mu}$ and $\vec{B}_0$ is given by:

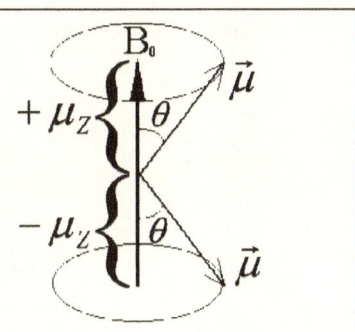

Fig. 1.2.2. Spin-$\frac{1}{2}$ nuclei in a static magnetic field, $\vec{B}_0$. $\vec{\mu}$ can only take on two possible orientations relative to $\vec{B}_0$, thus μ_z can either be parallel or anti-parallel to $\vec{B}_0$.

$$E = \vec{\mu} \bullet \vec{B}_0 = -\mu B_0 \cos(\theta) = \mu B_0 \frac{m_I}{[I(I+1)]^{\frac{1}{2}}}. \quad [1.2.4]$$

For a spin-$\frac{1}{2}$ system, there are only two possible orientations of $\vec{\mu}$, hence the z-component of the nuclear dipole moment, μ_z, can either be parallel or anti-parallel with respect to $\vec{B}_0$ (Fig. 1.2.2). The energy associated with each orientation is given by

NMR THEORY

$$E = \pm \mu_z B_0 = \pm \frac{\gamma \hbar B_0}{2}. \quad [1.2.5]$$

Consequently, the differences in energy between the two states would be

$$\Delta E = \frac{\gamma \hbar B_0}{2} - \left(-\frac{\gamma \hbar B_0}{2}\right) = \gamma \hbar B_0 = \frac{\gamma}{2\pi} h B_0. \quad [1.2.6]$$

Since energy is related to frequency, ν (Halliday and Resnick, 1988), by Planck's constant, h (Eq. [1.2.7])

$$\Delta E = E_a - E_b = h\nu, \quad [1.2.7]$$

Eqs. [1.2.6] and [1.2.7] can be equated to give

$$\frac{\gamma}{2\pi} h B_0 = h\nu. \quad [1.2.8]$$

Simplification of this equation yields

$$\omega = 2\pi\nu = \gamma B_0. \quad [1.2.9]$$

Eq. [1.2.9] is the quantum mechanical expression for the resonance condition and indicates that the frequency of electromagnetic radiation, which induces a transition between energy levels, is directly proportional to the magnetic field strength, B_0, by the gyromagnetic ratio, γ.

Up to this point, we have discussed the behavior of individual magnetic moments in a static magnetic field. However, in NMR, the signal arises from the vector sum of the individual $\vec{\mu}$. This distribution of individual magnetic moments amongst the various energy levels is described by the Boltzmann equation:

$$N_I = e^{-\frac{E_I}{kT}} \qquad [1.2.10]$$

where k is the Boltzmann constant ($1.38 \times 10^{-23} \frac{J}{K}$), and N_I is the number of spins at energy E_I and temperature T (in Kelvin). Since only population differences between energy states can be detected by NMR, and the population difference can be represented as

$$\Delta n = \frac{\Delta E}{2kT} \sum N_I = \frac{\gamma \hbar B_0}{2kT} \sum N_I, \qquad [1.2.11]$$

and we see that the total signal, M_0, can be represented as

$$M_0 = \frac{1}{2} \gamma \hbar (\Delta n) = \frac{\gamma^2 \hbar^2 B_0}{4kT} \sum N_I \qquad [1.2.12]$$

NMR THEORY

where $\sum N_I$ is the total number of spins in the system.

Another useful way to visualize the behavior of the nucleus is to consider $\vec{\mu}$ classically as a small magnet. If $\vec{\mu}$ is inclined at an angle with respect to $\vec{B}_0$ (Fig. 1.2.2), a torque is exerted on $\vec{\mu}$ which causes it to precess about $\vec{B}_0$. The precessional behavior of $\vec{\mu}$ is described by (Callaghan, 1991)

$$\frac{d\vec{\mu}}{dt} = \gamma\vec{\mu} \times \vec{B}_0. \qquad [1.2.13]$$

Since the net magnetization vector is the sum of the individual spins, the behavior of $\vec{M}_0$ can be described as the vector sum of the individual dipole moments

$$\sum \frac{d\vec{\mu}}{dt} = \sum \left(\gamma\vec{\mu} \times \vec{B}_0\right) \qquad [1.2.14]$$

or

$$\frac{d\vec{M}_0}{dt} = \sum \frac{d\vec{\mu}}{dt} = \sum \left(\gamma\vec{\mu} \times \vec{B}_0\right) = \gamma\vec{M}_0 \times \vec{B}_0. \qquad [1.2.15]$$

The solution (Callaghan, 1991; Halliday and Resnick, 1988) to Eq. [1.2.15] corresponds to a precession of the magnetization about the $\vec{B}_0$ field at a rate of

$$\omega_0 = \gamma B_0. \qquad [1.2.16]$$

Eq. [1.2.16] is known as the Larmor equation and is the fundamental equation describing the precessional behavior of spins in an external magnetic field.

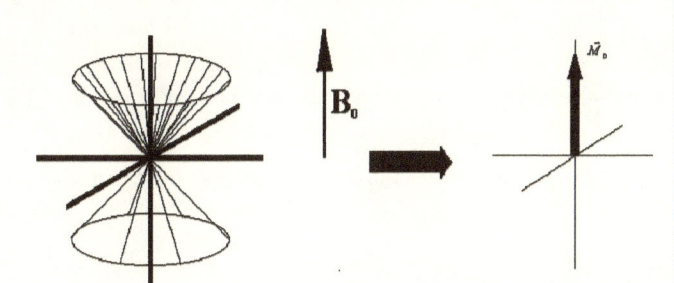

Fig. 1.2.3. The net magnetization. Individual magnetic moments sum to form the net magnetization vector, $\vec{M}_0$. Due to the random orientation of the spins about the z-axis, the transverse components of the spin ensemble sum to zero. Also, since more spins align parallel to the field than anti-parallel, the z-components of the spin ensemble sum to give the magnitude of the net magnetization vector, $\vec{M}_0$.

1.3. Rotating Frame of Reference and Radio-Frequency Pulses

Due to the precession of the net magnetization vector produced by placing the nuclei into the static magnetic field, it is convenient to consider a rotating frame of reference equal to that of the Larmor frequency of the nuclei under observation. In this reference frame, the precessing magnetization appears stationary and allows for simpler visualization of the behavior of the net magnetization. This coordinate system transformation allows us to rewrite the equation of motion of the net magnetization as

$$\left.\frac{d\vec{M}}{dt}\right|_{rotating} = \gamma \vec{M} \times \left(\vec{B}_0 + \frac{\vec{\omega}}{\gamma} \right). \qquad [1.3.1]$$

where $\frac{\vec{\omega}}{\gamma}$ represents the term from the rotating frame of reference. It can be seen from this equation that an exact cancellation of the influence of $\vec{B}_0$ can be achieved by setting

$$-\frac{\vec{\omega}}{\gamma} = \vec{B}_0, \qquad [1.3.2]$$

a condition which is satisfied when the frequency of the rotating frame of reference ($\vec{\omega}$) is equal to that of the Larmor frequency. Since the Larmor equation describes a precessional dependence of the net magnetization vector on the static magnetic field, the deconvolution of this precessional term depicts the magnetization as a stationary vector in the rotating frame of reference (i.e. $\left.\frac{d\vec{M}}{dt}\right|_{rotating} = 0$). Further, if the nuclei are irradiated with an oscillating radiofrequency (RF) magnetic field, $\vec{B}_1$, Eq. [1.3.1] can be modified to

$$\left.\frac{d\vec{M}}{dt}\right|_{rotating} = \gamma \vec{M} \times \left(\vec{B}_0 + \frac{\vec{\omega}}{\gamma} + \vec{B}_1 \right). \qquad [1.3.3]$$

Now if the frequency of the rotating frame of reference is set exactly equal to the Larmor frequency (Eq. [1.3.2]), the static term and the precessional term cancel and Eq. [1.3.3] simplifies to:

$$\left.\frac{d\vec{M}}{dt}\right|_{rotating} = \gamma \vec{M} \times \vec{B}_1 . \qquad [1.3.4]$$

Thus, from a frame of reference that is rotating at the Larmor frequency, an oscillating RF field of exactly the same frequency in a plane perpendicular to $\vec{B}_0$ produces a static $\vec{B}_1$ field in the rotating frame of reference that is orthonormal to $\vec{B}_0$. The spins will precess about $\vec{B}_1$ at a frequency given by

$$\omega_1 = -\gamma B_1 . \qquad [1.3.5]$$

If the duration of the oscillating RF field can be controlled, the precessional or tip angle of the net magnetization around the orthonormal axis can be calculated using

$$\theta_{tip} = \gamma B_1 t_{tip} \qquad [1.3.6]$$

where t_{tip} is the duration of the oscillating RF field and θ_{tip} is the tip or flip angle. By controlling the duration (t_{tip}) and/or the amplitude (B_1) of the RF excitation pulse, an arbitrary flip angle can be achieved.

1.4. The Free-Induction Decay (FID)

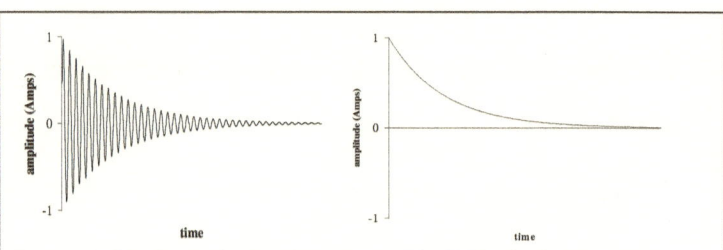

Fig. 1.4.1. The free-induction decay (FID). The FID (left) shows the time-varying current induced into a resonant circuit from the magnetization precessing in the transverse plane. The sinusoidal behavior of the FID is enveloped in a decaying exponential characterized by a time constant, T_2^*. The same FID as seen from the rotating frame of reference with a frequency exactly equal to that of the Larmor frequency (right) exhibits the behavior of a decaying exponential with no oscillating component.

Following RF excitation, the net magnetization vector precesses in the transverse plane and acts as a time-varying magnetic field oscillating at the Larmor frequency. According to Faraday's Law, this time-varying magnetic field can induce a current into a resonant circuit tuned to the oscillating frequency. This detected signal is referred to as the free-induction decay (FID) and is shown in Fig. 1.4.1.

Two important characteristics of the FID are the frequency component and the component of decay. Since the excited spins precess at their respective Larmor frequencies, and the system is observed from the rotating frame of reference, the frequency components within the FID are simply the differences between the rotating frame frequency and the Larmor frequencies. The exponential decay constant is a measure of how quickly the observed signal (or induced current) disappears.

Given this, the behavior of the magnitude of the transverse magnetization (and hence the FID) in the rotating frame of reference can be described by

$$M_{xy} = M_0 e^{-\frac{t}{T_2^*}} \qquad [1.4.2]$$

where M_{xy} is the observed signal intensity of the FID at time, t, and M_0 is the initial amplitude of the signal following a 90° RF pulse.

1.5. Fourier Transform and the Frequency Domain

The FID is collected in the time domain and gives the characteristic behavior of the signal as a function of time. However, since the measured signal is characterized by its respective Larmor frequency components, another mode with which the signal can be visualized is in the frequency domain. Although the information contained in the time domain signal is identical to that of the frequency domain, the extraction of complex frequency components is made simpler when viewed from a frequency standpoint.

The Fourier relation between the time domain signal, $f(t)$, and its frequency domain representation, $F(\omega)$, is given by (Peebles, 1987)

$$F(\omega) = \int_{-\infty}^{\infty} f(t) e^{-i\omega t} dt \qquad [1.5.1]$$

for the Fourier transform (FT), and

$$f(t) = \frac{1}{2\pi} \int_{-\infty}^{\infty} F(\omega) e^{i\omega t} d\omega \qquad [1.5.2]$$

for the inverse of the Fourier transform (IFT). Since the FT and the IFT are merely representations of each other in reciprocal domains (i.e. time v. frequency), a simple multiplication in the time domain would be equivalent to a convolution in the frequency domain, and *vice versa*. This is known as the convolution theorem (Peebles, 1987). Several important Fourier pairs commonly used in NMR are shown in Fig. 1.5.1.

Since the FID takes the shape of $\sin(\omega_0) e^{-\frac{t}{T_2^*}}$, it is important to know the frequency domain representation of this exponentially decaying sine wave. The FT of the FID would be a Lorentzian centered around ω_0 with a full-width at half-maximum (FWHM) height of

$$f_{FWHM} = \frac{1}{\pi T_2^*} \qquad [1.5.3]$$

Another function of importance in NMR is the rectangular wave. A rectangular wave symmetrically disposed about the origin in the frequency domain corresponds to a $\frac{\sin(x)}{x}$ function (also called a sinc function) in the time domain. Recalling the convolution theorem, we see that enveloping a sine wave with a sinc function (multiplication of a sinc function and a sine wave) acts to shift the rectangular wave in the frequency domain (convolution of a rectangular wave with a $\delta(\omega - \omega_0)$ function where ω_0 is the frequency of the sine wave contained in the rectangular envelope) (Fig. 1.5.3).

NMR THEORY

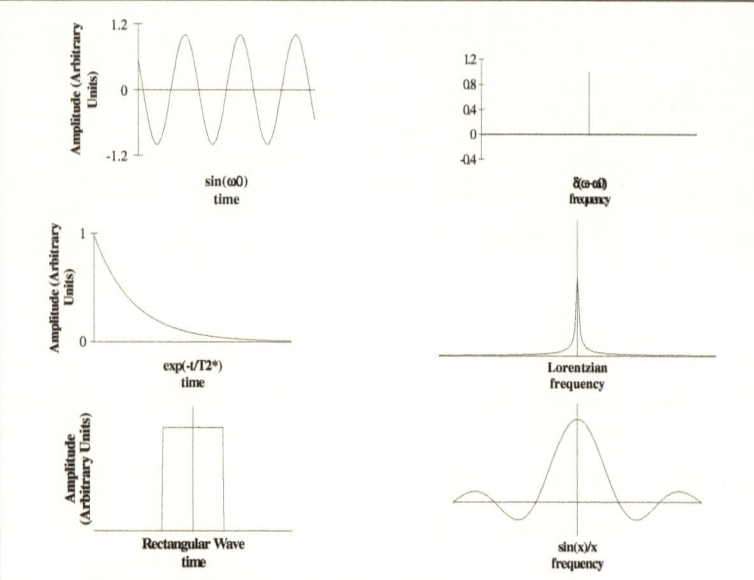

Fig. 1.5.1. Fourier pairs commonly used in NMR.
- The FT of a sine wave is a δ-function at the frequency of the sine wave.
- The FT of a decaying exponential is a Lorentzian line at zero frequency whose full width at half-maximum height is $\frac{1}{\pi T_2^*}$.
- The FT of a rectangular wave from $\pm t$ is a $\frac{\sin(x)}{x}$ function with a lobe width of $\frac{1}{t}$.

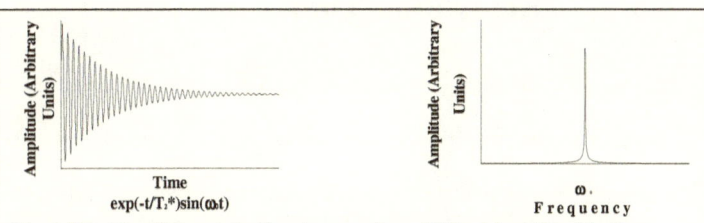

Fig. 1.5.2. The FID and the Fourier transform (FT) of the FID. The line shape that results from the FT of the FID is characterized by a Lorentzian. The FID displays information from the time domain while the FT of the FID displays frequency components. In this example, the center frequency of the Lorentzian is the frequency of the enveloped sine wave, and the full width at half-maximum height (FWHM) of the Lorentzian is proportional to $\dfrac{1}{\pi T_2^*}$.

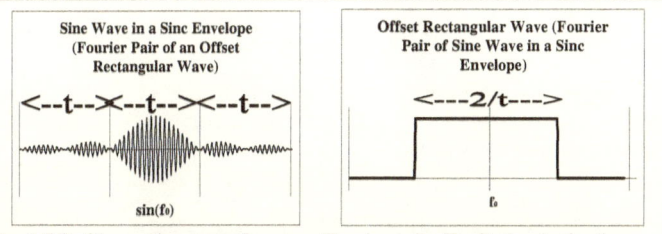

Fig. 1.5.3. Sinc and rectangular wave Fourier pair. Excitation in the time domain *via* a sinc functional form is analogous to excitation of a rectangular wave in the frequency domain. Enveloping a sine wave with frequency, f_0, in a sinc envelope in the time domain shifts the rectangular wave in the frequency domain to a position centered around f_0.

1.6. Relaxation Time Constants

Following RF excitation, the spin system recovers back to Boltzmann equilibrium by two relaxation processes. Here the processes are briefly discussed as well as methods for measuring the associated relaxation time constants. A more detailed explanation of these relaxation processes can be found in writings by Bloch (1946a, 1946b) or Abragam (1961).

1.6.1. T_1 Relaxation (Spin-Lattice or Longitudinal Relaxation)

Recalling that the NMR signal is generated from the vector sum of parallel and anti-parallel spins (Section 1.2, Fig. 1.2.3), we see that at Boltzmann equilibrium, more spins are aligned parallel to the external magnetic field than anti-parallel. A 180° inversion pulse deposits energy into the system causing a transition of spins from the lower energy state, E_1, to the higher energy state, E_2 (Fig. 1.6.1.1). Immediately following this inversion pulse, the spin system begins to evolve back to thermal equilibrium by releasing its deposited energy to the surroundings (the "lattice"). This spin-lattice relaxation is governed by a time constant, T_1, and the differential equation describing this phenomena in the rotating frame of reference (Abragam, 1961) can be written as

$$\frac{dM_z}{dt} = -\frac{1}{T_1}(M_z - M_0). \qquad [1.6.1.1]$$

Since the spins are randomly ordered around the z-axis, the x- and y-components of M cancel and M_z is the only nonzero component. Therefore, when spins are inverted by a 180° RF pulse, only the magnitude of M_z is affected. Similarly, when the spins relax back to Boltzmann equilibrium, only M_z is affected.

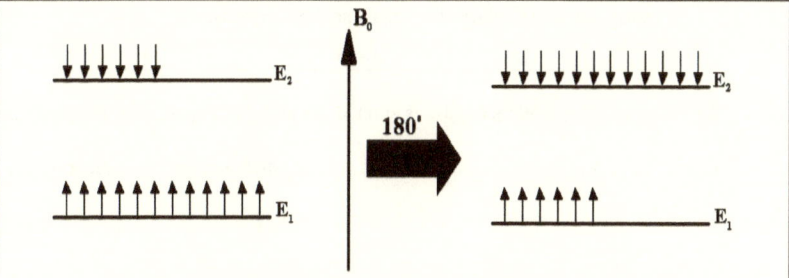

Fig. 1.6.1.1. Inversion of spins upon application of a 180° pulse. Absorption of energy causes an excitation of spins from the lower energy state, E_1, to the higher energy state, E_2. Once excited, these excited spins relax back to ground state (E_1) as energy is released to the lattice. Since only the z-component of relaxation is affected by T_1 relaxation, the spin states in E_1 and E_2 represent the parallel and anti-parallel μ_z.

The solution of this differential equation is

$$M_z(t) = M_0 \left(1 - e^{-\frac{t}{T_1}}\right) \qquad [1.6.1.2]$$

NMR THEORY

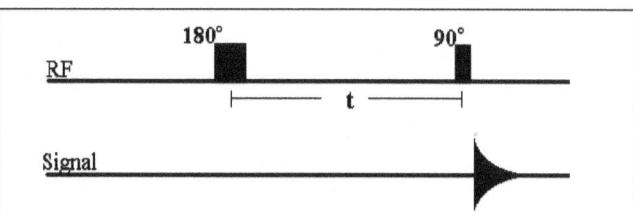

Fig. 1.6.1.2. Inversion-recovery (IR) pulse sequence. A 180° RF pulse is followed by an evolution time (*t*). During this time, the inverted magnetization recovers along the longitudinal or z-axis. The 90° RF pulse tips the longitudinal magnetization at time *t* into the transverse plane, where the signal can then be measured.

for the case where a 90° RF pulse is applied. In an NMR experiment, the time *t* is usually referred to as the repetition time (*TR*, the time at which the spins are successively excited by the 90° RF pulse).

One way to measure the T_1-relaxation time constant is by using a pulse sequence known as an inversion-recovery (IR) pulse sequence (Fig. 1.6.1.2). In this method, a 180° RF pulse is applied to the spins to invert the net magnetization vector. Following inversion, as the spins recovers to Boltzmann equilibrium, a 90° RF pulse is applied at various time intervals (*TI*), thus tipping the magnetization into the transverse plane where the signal can be measured. The measured signal will be dependent on *TI* (Fig. 1.6.1.3) according to:

$$M_{xy} = M_0 \left(1 - 2e^{-\frac{TI}{T_1}}\right). \qquad [1.6.1.3]$$

NMR THEORY

If the inversion pulse is not exactly 180°, then the above equation will take on the form

$$M_{xy} = A\left(1 - Be^{-\frac{t}{T_1}}\right) \qquad [1.6.1.4]$$

where A is some fraction of M_0 and B (a number less than two) gives some indication of the inversion efficiency. The measured data can then be fitted for T_1 using Eq. [1.6.1.4].

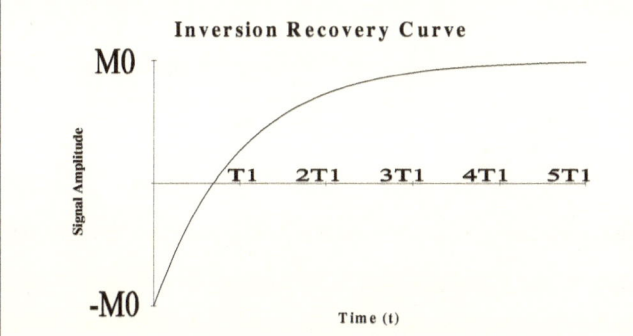

Fig. 1.6.1.3. Inversion-recovery (IR) curve. The behavior of the observable signal for an IR pulse sequence is plotted. The signal is dependent on the evolution time (t) according to

$$M_{xy} = M_0\left(1 - 2e^{-\frac{t}{T_1}}\right).$$

After $t = 5T_1$, the longitudinal magnetization has almost fully recovered to the Boltzmann equilibrium.

1.6.2. T_2 Relaxation (Spin-Spin Relaxation)

While the T_1 relaxation-time constant describes the recovery of the spin system to the Boltzmann equilibrium following RF excitation, the T_2 relaxation time is an indication of how quickly the observable signal disappears. The observable signal is comprised of the vector sum of individual spins, and it is the coherent precession of these individual spins that generates the NMR signal.

The resonant condition between spins of the same frequency allows for energy exchange between these spins and is responsible for spin-spin relaxation. Intra- and intermolecular interactions cause the local magnetic field around the spins to fluctuate causing slight modulations in ω_0 experienced by these spins. This fluctuation produces a gradual loss of phase coherence in the net transverse magnetization as the spins exchange energy and leads to an attenuation of the transverse magnetization.

There are several possible causes for a loss of phase coherence. The modulation of the local magnetic fields due to molecular interactions (T_2) is one of the causes, but magnetic field inhomogeneities and other factors can also contribute to the dephasing of the transverse magnetization. Nuclear spins located in different parts of the sample experience slightly different B_0 fields and precess at different Larmor frequencies, thus causing a loss of phase coherence. For this reason, the attenuation of the signal (or FID) is not only affected by the inherent T_2, but also diffusion, magnetic susceptibility, and inhomogeneity effects. All of these factors contribute to a time constant known as T_2^* according to

$$\frac{1}{T_2^*} = \frac{1}{T_{2_{intrinsic}}} + \frac{1}{T_{2_{diffusion}}} + \frac{1}{T_{2_{susceptibility}}} + \frac{1}{T_{2_{inhomogeneities}}}.$$ [1.6.2.1]

For the purposes of many NMR experiments, it would be advantageous to recover the signal loss due to susceptibility and inhomogeneity effects. A method to reverse the time evolution of spins was presented in a seminal paper by Erwin Hahn (1950). It was shown that a 180° RF pulse applied τ seconds after an initial 90° excitation RF pulse would cause the phase reversal necessary for refocusing the transverse magnetization at time 2τ (Figs. 1.6.2.1 and 1.6.2.2). The refocused echo is known as the Hahn spin echo, and the attenuation of the signal at 2τ becomes a function only of the intrinsic T_2 relaxation time of the nuclei. The equation of motion for the transverse component of magnetization, M_{xy}, in the rotating frame of reference is given by

$$\frac{dM_{xy}}{dt} = -\frac{1}{T_2} M_{xy}$$ [1.6.2.2]

which has the solution

NMR THEORY

$$M_{xy} = M_0 e^{-\frac{t}{T_2}} \quad\quad [1.6.2.3]$$

where t is the echo time ($t = 2\tau$), or the time between the first RF excitation and acquisition. M_0 is the magnitude of the transverse magnetization immediately following a 90° RF pulse applied to an equilibrium spin system.

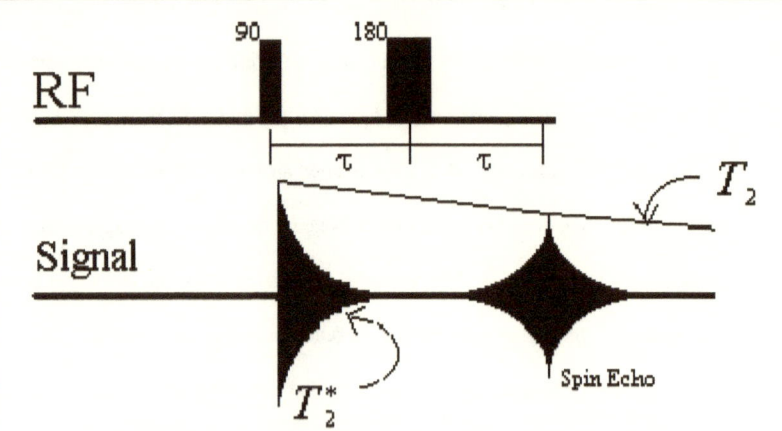

Fig. 1.6.2.1. Hahn spin echo (SE) pulse sequence. Spins dephase during the time interval (τ) after the 90° RF excitation by T_2^*. The 180° RF pulse causes a phase reversal allowing the of spins to refocus at time 2τ. This refocused echo is known as the Hahn spin echo.

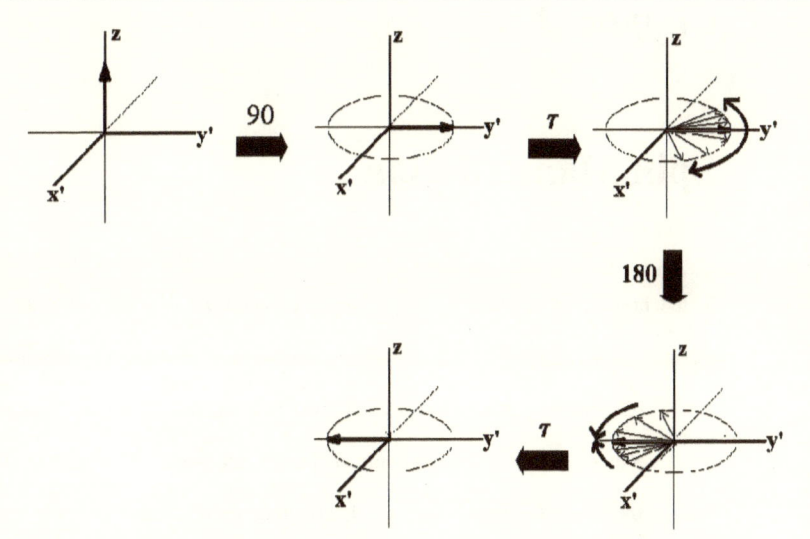

Fig. 1.6.2.2. Spin behavior during Hahn spin echo pulse sequence. The initial 90° excitation pulse tips the net magnetization into the transverse plane. During the time interval between the 90° and 180° RF pulse (τ), these spins dephase due to magnetic field inhomogeneity and other effects. A 180° RF pulse flips the spins around an axis causing a phase reversal. The spins refocus at time τ after the 180° RF pulse and the echo that forms is known as the Hahn spin echo.

Chapter 2

Spin Manipulation

In the previous chapter, the topic of spin excitation was discussed from a quantum mechanical standpoint. In this chapter, the various methods of manipulating the macroscopic net magnetization vector, M, will be discussed. There are essentially three major components to signals: amplitude, frequency, and phase. This chapter will briefly discuss methods of manipulating and interpreting signals based on their amplitudes, frequencies, and phases.

2.1. The Chemical Shift and Binomial Pulses

The nuclei of atoms are surrounded by electron clouds that act to shield the nuclei from the external magnetic field. Due to the perturbation in the magnetic field experienced by these shielded nuclei, their Larmor frequencies are shifted in proportion to the perturbation in the chemical environment caused by the magnetic shielding. This shielding effect alters the magnetic field (Noggle and Schirmer, 1971) so that the field experienced by the nuclei is

$$B_i = B_0(1-\sigma_i) \qquad [2.1.1]$$

and, consequently, the Larmor frequency is

$$\omega_i = \gamma B_0(1-\sigma_i) \qquad [2.1.2]$$

where B_i is the effective magnetic field seen by the nuclei, B_0 is the strength of the main magnetic field, σ_i is the chemical shielding term for nucleus i, and ω_i is its precessional frequency in rad/sec. In general, chemical shielding is very small compared to the main magnetic field and is referred to as the chemical shift.

Fig. 2.1.1. 1-3-3-1 binomial pulse. The τ intervals are determined by the chemical shift difference (Δf, in absolute frequency) between the desired and undesired NMR peaks. These intervals between pulses (τ_1, τ_2, and τ_3) are exactly $(2\Delta f)^{-1}$ seconds. An example of the spin behavior for a 1-3-3-1 pulse train is given in Fig. 2.1.2.

In NMR experiments, it is sometimes desirable to suppress the signal from a particular chemical species in the sample. If the chemical shift of the unwanted spins is sufficiently different from that of the desired signal in the sample, then binomial RF pulses may be used to suppress the unwanted signal (Hore, 1983). One specific example of this is the 1-3-3-1 binomial pulse sequence and is illustrated in Figs. 2.1.1 and 2.1.2. The name of the 1-3-3-1 pulse sequence comes from

SPIN MANIPULATION

the fact that the ratios of the individual pulse durations are 1:3:3:1 (Fig. 2.1.1) while the sum of these pulses determines the flip angle. Thus, for a flip angle of 90°, the individual pulses would be 11.25°-33.75°-33.75°-11.25°. τ is the time interval between the center of one RF pulse to the next and is calculated as $(2\Delta f)^{-1}$, where Δf is the chemical-shift difference in absolute frequency units (i.e. the frequency separation between the water and fat NMR peaks).

If the RF transmitter is set to the Larmor frequency of the desired signal (i.e. water), then

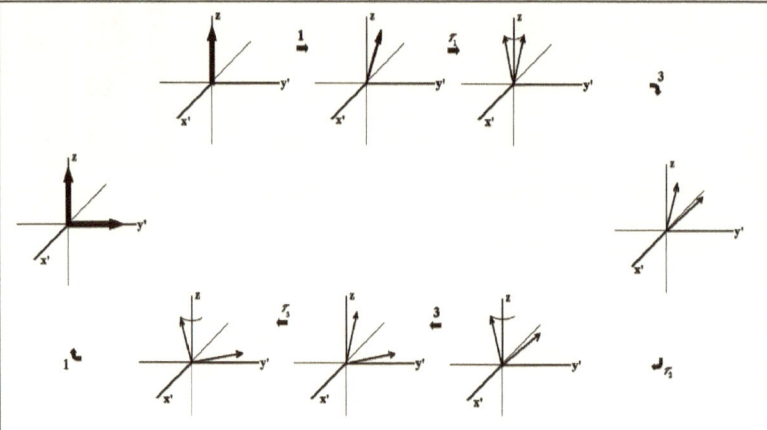

Fig. 2.1.2. Illustration of a 1-3-3-1 chemical-shift-selective binomial pulse sequence on the net magnetizations of water and fat. After the specific order of RF pulses (1, 3, 3, and 1) and time intervals (tau1, tau2, and tau3) have been applied, the desired water signal is tipped into the transverse plane while the undesired fat signal is maintained in the longitudinal axis. The time interval from the center of one RF pulse to the next is exactly $(2\Delta f)^{-1}$ seconds where Δf is the chemical-shift difference (in absolute frequency units) between water and fat. In this example, the tip angle is 90° leading to 1-3-3-1 pulses of 11.25°-33.75°-33.75°-11.25°.

upon application of the initial 11.25° pulse and a delay of τ, both spin systems (i.e. water and fat) will be 180° out of phase with each other (refer to Fig. 2.1.2). At this point, the application of a 33.75° pulse would lead to a total of 45° and 22.5° flip angle for the resonant and off-resonant spins, respectively. After another time delay of τ, the two systems will once again be 180° out of phase. We see that by the end of the 1-3-3-1 pulse sequence, the effective flip angle experienced by the spin system on resonance is 90° while the spin system off resonance experiences a 0° flip angle. Therefore, none of the spins of the unwanted species will have a net excitation. Binomial pulses have been successfully employed in NMR spectroscopy for solvent suppression.

2.2. Spin Echoes (SE), Stimulated Echoes (STE), and Spurious Echoes

Hahn (1950), in his early works, has shown that any two RF pulses can form an echo. In Section 1.6.2, the spin-echo pulse sequence was discussed as a method for refocusing the dephased spins associated with magnetic-field-inhomogeneity effects. The spin-echo pulse sequence is an example of how two RF pulses (90° and 180°) can form an echo. Recalling that the signal attenuation is a function of echo time, we see that the signal-to-noise becomes progressively worse with longer echo times.

In some NMR experiments, it becomes necessary to observe the signal long after the initial RF excitation. This becomes a problem when the T_2 of the sample is much shorter than the evolution time of the spins necessary to observe such an event. Since the signal attenuates as

$$M_{xy} \propto e^{-\frac{t}{T_2}},$$
[2.2.1]

this limits the use of the spin echo sequence to echo times ranging on the order of T_2. For this reason, it would be convenient to be able to store the coherence of spins over a longer time interval. Since the T_1 is usually much longer than T_2 in biological systems, one possibility is to manipulate the spins such that the net magnetization is restored to the longitudinal axis during part of the evolution time.

The stimulated echo (STE) pulse sequence (Fig. 2.2.1) shows an example of how this can be achieved (Hahn, 1950). The second 90° RF pulse in the STE pulse sequence stores half of the magnetization along the longitudinal axis until the next RF excitation pulse is applied. During the evolution time (TM) the magnetization stored along the longitudinal axis is only affected by the T_1 relaxation. The crusher gradient applied during the TM period completely dephases the half of the magnetization that remains in the transverse plane following the second 90° RF pulse.

This dephasing of spins results in a reduction of signal by a factor of two for the STE compared to a SE sequence of equal echo time (TE). One point to note is that an echo is formed from each pair of RF pulses, and, depending on the RF pulse timing, it is possible

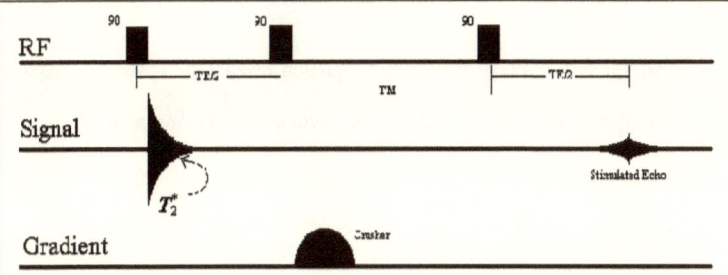

Fig. 2.2.1. Stimulated-echo (STE) pulse sequence. After the initial 90° pulse, an FID is formed. The second 90° pulse stores half of the magnetization along the longitudinal axis while the third 90° pulse restores the magnetization to the transverse plane and causes a stimulated echo to form. Note the "crusher" gradient between the second and third 90° RF pulses used to destroy the unwanted half of the transverse magnetization that remains following the second 90° RF pulse. Without this crusher, four possible echoes could be detected – echoes formed from (1) the first-second 90° combination; (2) the first-third 90° combination; (3) the second-third 90° combination; and (4) the stimulated echo.

to generate spurious echoes that interfere with the desired stimulated echo. Since only the spins stored along the longitudinal axis are unaffected by gradients during TM, spurious echoes due to the other combinations of RF pulses can be effectively eliminated by applying a large crusher gradient during the TM period.

SPIN MANIPULATION 31

2.3. Magnetic Resonance Imaging

2.3.1. Effect of Magnetic Field Gradients on Spins

A careful inspection of the Larmor equation reveals the basis for imaging. Since the precessional frequency is directly proportional to the magnetic field strength by the Larmor equation,

$$\omega_0 = \gamma B_0 \qquad [2.3.1]$$

we see that in a uniform magnetic field (B_0), all of the spins in a sample will precess at the same rotational frequency (ω_0). By applying a linear magnetic field gradient in addition to the static magnetic field, it is possible to manipulate the spins in

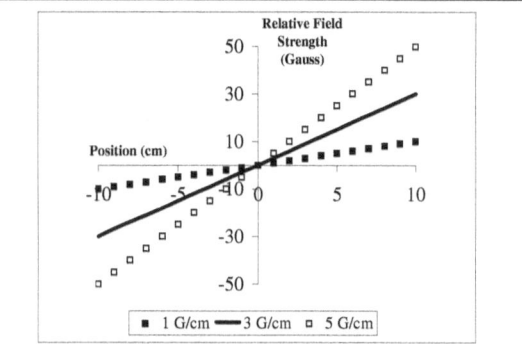

Fig. 2.3.1.1. Relative field strength as a function of spatial position. With the application of a linear magnetic field gradient, spins in different spatial locations experience different magnetic fields. This causes the Larmor frequencies of the precessing spins to be directly proportional to the spatial location of the spins.

such a manner so as to make them precess at a rate proportional to their respective locations. Fig. 2.3.1.1 shows the relative field strength as a function of spatial location for different applied linear gradients. Since the spatial location is linearly related to the relative field strength, it must be (according to the Larmor equation) that the frequency is linearly related to position. Strategic use of controlled, linear, magnetic-field gradients allow for manipulation of the frequency and phase components of the signal. These frequency and phase components are used for spatial encoding in magnetic resonance imaging and will be discussed in the following sections.

2.3.2. Frequency Encoding

Recalling from the Larmor equation that the precessional frequency is linearly related to the field strength, we see that if the magnetic field changes by $+g$, the precessional frequency will be altered as

$$\omega = \gamma(B_0 + g). \quad\quad\quad [2.3.2.1]$$

By adding a controlled, linear, magnetic-field gradient to the static magnetic field, we see that the strength of the magnetic field can be manipulated as a function of spatial location (Fig. 2.3.1.1). Since the field strength is directly proportional to position for linear gradients, and since frequency is directly proportional field strength, if the signal is acquired in the presence of a known gradient, it is possible to back calculate the spatial location of spins from the frequency information using

$$Location(cm) = \frac{frequency(Hz)}{\frac{\gamma}{2\pi}\left(\frac{Hz}{G}\right) \times gradient\left(\frac{G}{cm}\right)}.$$ [2.3.2.1]

The applied linear gradient during signal acquisition is known as the frequency-encoding gradient or readout gradient, and the frequency component of the acquired signal constitutes one of the parameters of the raw data used for magnetic resonance imaging (MRI).

2.3.3. Slice Selection

Often, for imaging or localized spectroscopy, it becomes necessary to excite only a certain portion of the sample under observation. One of the basic methods for localized excitation is *via* slice-selective RF pulses (also known as "soft" pulses). By taking advantage of the Fourier relationship between time and frequency, shaped RF pulses in the time domain can selectively excite specific slice profiles in the frequency domain. Most commonly, a rectangular slice profile is used. Recall that the FT of a sinc function in the time domain yields a rectangular function in the frequency domain, and that the multiplication of a sinc function with a sine wave produces an offset square wave (Section 1.5). Thus, in order to excite a rectangular slice profile with a frequency bandwidth of Δf Hz, an RF pulse enveloped in a $\text{sinc}\left(\dfrac{1}{2\Delta f}t\right)$ function would have to be applied (see Fig. 2.3.3.1). If the underlying RF wave has a frequency, f_0, the excited frequency bandwidth would be $f_0 \pm \dfrac{\Delta f}{2}$.

One point of concern is that in order to excite a perfectly rectangular slice profile, a sinc pulse with an infinite number of side lobes would be necessary. Due to practical limitations, this is not possible and a compromise is made between the number of side lobes and the ideality of the rectangular profile. A reasonable approximation of the rectangular profile can be realized by truncating the sinc pulse to three or five side lobes. The Fourier transform of a truncated sinc pulse is a trapezoidal function with ringing artifacts. These ringing artifacts appear as decaying ripples on the rising and falling edges of the trapezoidal waveform due to the truncation of high frequency components of the sinc pulse. Truncation artifacts are minimal for most imaging experiments and can often be neglected.

A greater degree of versatility in slice selection is possible through the use of linear magnetic field gradients. According to the Larmor equation, the frequency, ω, of nuclei in the presence of a magnetic field gradient, g, superimposed onto the static magnetic field, B_0, is

$$\omega = \gamma(B_0 + g). \qquad [2.3.3.1]$$

SPIN MANIPULATION

or

$$\Delta\omega = \gamma(\Delta g). \qquad [2.3.3.2]$$

Since the slice thickness is directly proportional to the frequency profile, and since the frequency profile is directly proportional to the gradient field strength (Eq. [2.3.3.2]), it is possible to vary the slice thickness by simply varying the gradient strength according to

$$Slice\_Thickness(cm) = \frac{frequency(Hz)}{\frac{\gamma}{2\pi}\left(\frac{Hz}{G}\right) \times gradient\left(\frac{G}{cm}\right)}. \qquad [2.3.3.3]$$

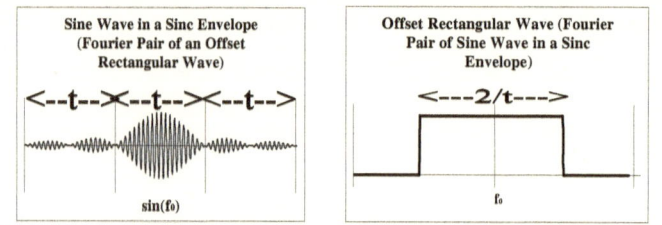

Fig. 2.3.3.1. Sinc and rectangular wave Fourier pair. Excitation in the time domain *via* a sinc functional form is analogous to excitation of a rectangular wave in the frequency domain. Enveloping a sine wave with frequency, f_0, in a sinc envelope in the time domain shifts the rectangular wave in the frequency domain to a position centered around f_0.

2.3.4. Phase Encoding

Since the amplitude of the transverse magnetization offers intensity information and the frequency component gives positional information in one dimension (Section 2.3.2), only the phase component of the signal is available for encoding information in the remaining direction.

When a gradient is applied, the spins precess at a rate proportional to the gradient strength, hence, a rate proportional to the spatial location of the nuclei. When the gradient is turned off, all of the nuclei will precess at the Larmor frequency of the B_0 field but will now have accumulated phase shifts that are a function of the strength and duration of the gradient. Knowledge of the phase accumulation can reveal spatial information.

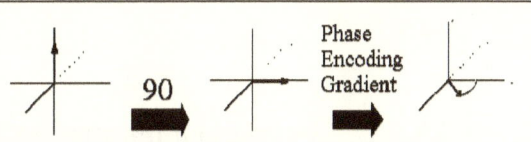

Figure 2.3.4.1. Phase encoding. After the initial excitation of M_0 into the transverse plane, a gradient applied for a finite duration, t_{phase}, imparts a phase, Φ, to M_{xy} according to

$$\Phi = \int_0^{t_{phase}} \left[\frac{\gamma}{2\pi} G_y(t_{phase}) y \right] dt_{phase}$$

where $G_y(t_{phase})$ is the gradient wave form and y is the location of the spins.

The phase information, Φ, contained in the NMR signal is related to the gradient waveform function, $G_y(t_{phase})$, the duration of the gradient pulse, t_{phase}, and the spatial position along the gradient axis (e.g., y) according to Eq. [2.3.4.1].

$$\Phi = \int_0^{t_{phase}} \left[\frac{\gamma}{2\pi} G_y(t_{phase}) y \right] dt_{phase} \qquad [2.3.4.1]$$

Thus, when a gradient is applied, a phase will be imparted to each spin that will reflect its spatial location along the phase-encoding-gradient axis. A schematic of this is shown in Fig. 2.3.4.1. This information is used in part to reconstruct the magnetic resonance image (MRI).

2.3.5. *k*-space and Imaging

Since a Fourier relationship exists between the time domain and the frequency domain, and the frequency of spins in the presence of a linear gradient is directly proportional to the position of those spins, it must be that the positions of the spins have a Fourier relation to their frequency. This inverse space, or *k*-space, vector (Mansfield and Grannell, 1975) is given by

$$\vec{k} = \frac{1}{2\pi} \gamma \vec{G} t \qquad [2.3.5.1]$$

where $\vec{G}$ is a linear gradient vector, t is the time, and γ is the gyromagnetic ratio of the nucleus. Since k-space is the Fourier analogue of real space, data acquired in k-space can be reconstructed into an image by direct Fourier transformation. The following gives a brief description of how data from an excited slice can be acquired using conventional two-dimensional Fourier imaging techniques and echo-planar Fourier imaging techniques.

2.3.5.1. Two-Dimensional Fourier Imaging

The conventional method for sampling k-space is by using frequency- and phase- encoding gradients. The frequency-encoding gradient encodes spatial information using the frequency component of the signal while the phase-encoding gradient serves to encode spatial information using the phase component of the acquired signal. Since the k-space vector is directly proportional to the gradient vector, incrementing the

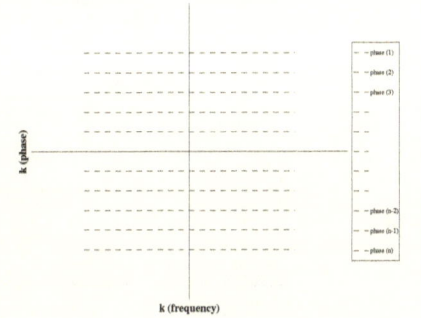

Fig. 2.3.5.1. Conventional two-dimensional Fourier imaging k-space raster. The phase-encoding gradient moves the k-space vector in one direction (vertical) while the frequency-encoding gradient sweeps across k-space (horizontal). The systematic sweeping of k-space using phase- and frequency-encoding gradients allows for a complete sampling of two-dimensional k-space.

phase-encoding gradient serves to move the *k*-space vector along one of the Cartesian directions of *k*-space. From here, if the frequency-encoding gradient is applied in a direction orthogonal to the phase-encoding direction during signal acquisition, information from that line of *k*-space is obtained (Fig. 2.3.5.1). By repeating this process for different phase-encoding gradient strengths, two-dimensional *k*-space can be systematically sampled (Figs. 2.3.5.1 and 2.3.5.2).

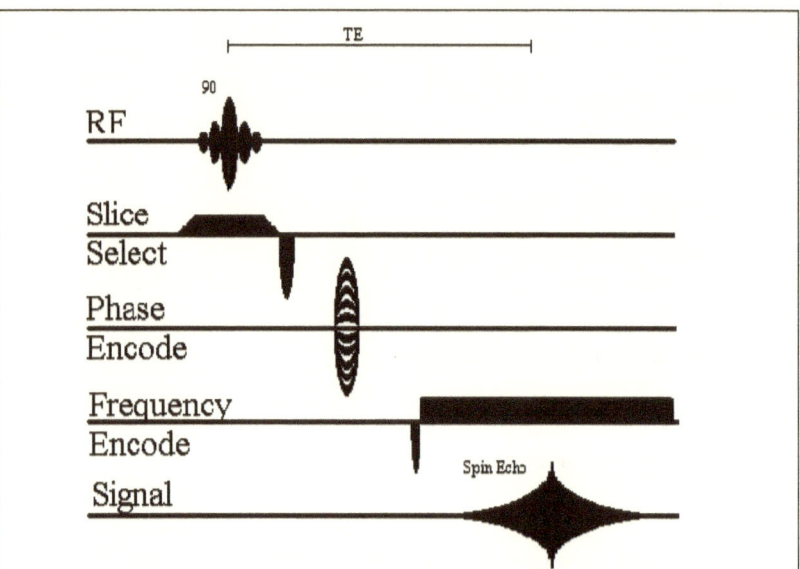

Fig. 2.3.5.2. Gradient-recalled echo (GRE) MRI pulse sequence. A sinc waveform is applied so as to excite a rectangular slice profile. The slice-select gradient determines the excitation slice thickness. The phase-encoding gradient imparts a phase to the excited spins and the frequency-encoding gradient allows for spatial encoding as a function of frequency. The *k*-space raster is systematically sampled by sequentially incrementing the phase-encode gradient strength and acquiring the signal in the presence of the frequency-encoding gradient. Image reconstruction is done by performing a two-dimensional Fourier transform of the acquired data.

The signal acquired from such a sampling of *k*-space can be represented as

$$S(k_x, k_y) = \int\limits_{-\infty}^{\infty}\int\limits_{-\infty}^{\infty} \rho(x, y) e^{i2\pi(k_x x + k_y y)} dx dy \qquad [2.3.5.2]$$

where x is the distance along the frequency-encoding direction k_x, y is the distance along the phase-encoding direction k_y, and $\rho(x, y)$ is the spin density function of the excited slice. Once the data has been acquired in this fashion, the image can be reconstructed by performing a two-dimensional Fourier transformation of the two-dimensional *k*-space data.

2.3.5.2. Saw-Tooth Echo-Planar Imaging

A careful breakdown of the frequency- and phase-encoding steps reveals a rectilinear sampling of the *k*-space. A major disadvantage of the conventional sampling of *k*-space is poor time efficiency. In a technique known as echo-planar imaging (EPI), the *k*-space data for a complete image can be acquired in a single acquisition.

By applying a gradient, called a broadening gradient (in place of a phase-encoding gradient), simultaneously with an oscillating gradient, two-dimensional *k*-space can be sampled in a single acquisition (Mansfield, 1977). The broadening gradient acts to sample *k*-space in one dimension while the oscillating gradient allows for a sweeping of *k*-space in the other dimension. Fig. 2.3.5.3 shows a typical SE-EPI pulse sequence

SPIN MANIPULATION

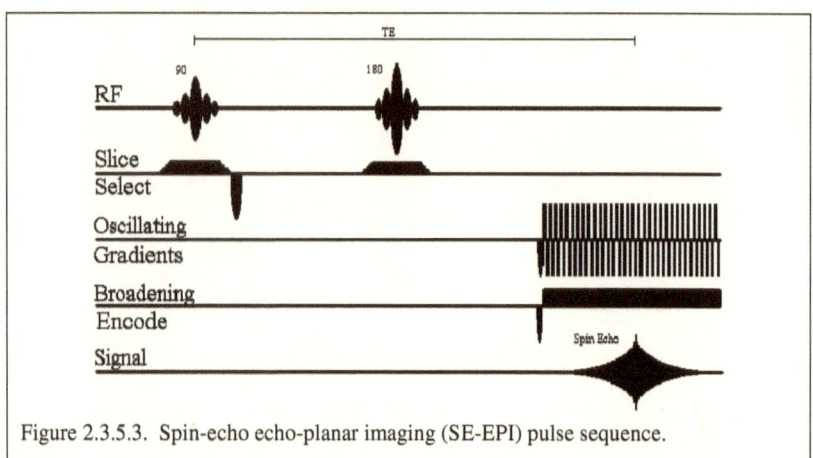

Figure 2.3.5.3. Spin-echo echo-planar imaging (SE-EPI) pulse sequence.

while Fig. 2.3.5.4 illustrates the *k*-space trajectory during the echo-planar detection. Since the broadening gradient and the oscillating gradient are concurrently applied, the sampling of *k*-space is not completely rectilinear, but rather, has a saw-tooth pattern. Since the acquisition has a finite duration, only the echoes formed at the center of the acquisition window will be T_2-weighted while the remaining gradient echoes will have some T_2^* contributions. For all practical purposes, large gradients can compensate for inhomogeneous field effects and short acquisition times can reduce T_2^* effects.

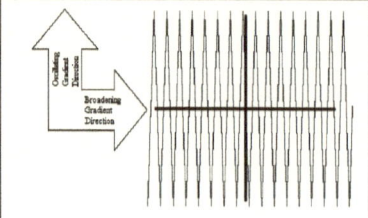

Figure 2.3.5.4. Saw-tooth *k*-space trajectory for echo-planar imaging. The broadening gradient moves the wave vector linearly in one dimension while the oscillating gradient causes the wave vector to alternate back and forth in the other dimension.

Since the oscillating gradients cause the sweeping of *k*-space to alternate from positive to negative (or vice versa) at every other sweep, it becomes necessary to reverse the data in every alternate sweep of the *k*-space raster before processing the EPI data. Once the data has been pre-processed correctly, the procedure for image reconstruction becomes identical to that of conventional two-dimensional Fourier imaging.

2.4. Diffusion

2.4.1. The Diffusion Process and Stochastic Modeling

When a concentration gradient, ∇C, exists between two compartments, a macroscopic flux between the two compartments can be observed according to Fick's law:

$$\vec{J} = -D\nabla C \qquad [2.4.1.1]$$

where $\vec{J}$ is the flux density, and D is the proportionality constant between the flux and the concentration gradient. Combining Fick's law with the equation for conservation of mass

$$\frac{\partial C}{\partial t} = -\nabla \bullet \vec{J}. \qquad [2.4.1.2]$$

we obtain the equation for diffusion

$$\frac{\partial C}{\partial t} = -\nabla \bullet \vec{J} = \nabla \bullet (D\nabla C). \qquad [2.4.1.3]$$

If we assume D to be constant, then Eq. [2.4.1.3] can be re-written as

$$\frac{1}{D}\frac{\partial C}{\partial t} = \nabla^2 C. \qquad [2.4.1.4]$$

For the case of constant diffusivity in an unbounded medium, the solution to the diffusion equation given a Dirac delta function as its input, $C(\vec{r},0) = \delta(\vec{r}-\vec{r}_0)$, is (LeBihan and Basser, 1995)

$$C(\vec{r},t) = \left(\frac{1}{\sqrt{4\pi Dt}}\right)^3 e^{-\left(\frac{(\vec{r}-\vec{r}_0)\bullet(\vec{r}-\vec{r}_0)}{4Dt}\right)} \qquad [2.4.1.5]$$

where $\vec{r}-\vec{r}_0$ is the displacement of the molecule during time t. In essence, spins initially distributed as a delta function ($\delta(\vec{r}-\vec{r}_0)$, i.e. a point source) will evolve over time to a Gaussian distribution characterized by D and t (Price, 1997). From here, we see that the mean-square displacement ($\langle x^2 \rangle = \langle (\vec{r}-\vec{r}_0)\bullet(\vec{r}-\vec{r}_0) \rangle$) of molecules is related to the diffusion constant, D, and time, t, by

$$\langle x^2 \rangle = 2nDt \qquad [2.4.1.6]$$

where n is the number of dimensions in which the molecules diffuse (one, two, or three dimensions). Thus, for translational diffusion in one dimension, Eq. [2.4.1.6] (known as the Einstein equation) becomes

$$\langle x^2 \rangle = 2Dt. \qquad [2.4.1.7]$$

Thus, information on the translational displacement of molecules over time allows us to calculate D.

2.4.2. Measuring Diffusion with Pulsed-Field Gradient NMR

In NMR experiments, diffusion is measured indirectly by measuring the mean-squared displacement of the sample's molecules. Typically, this is done using a Stejskal-Tanner pulsed-field gradient spin-echo (PFGSE, Fig. 2.4.2.1) or stimulated echo (PFGSTE) pulse sequence.

Previously, it was shown that positional information could be encoded by careful manipulation of the magnetic field (Section 2.3.4). Displacement (change in position) information can be encoded in a similar fashion.

Diffusion-encoding gradients operate in exactly the same way as phase-encoding gradients, by imparting a phase to the spins proportional to the area under the gradient waveform and the spatial location of spins. Recalling the phase-encode equation

$$\Phi = \int_0^\delta \left[\frac{\gamma}{2\pi} G_x(\delta) x \right] d\delta, \qquad [2.4.2.1]$$

we see that the phase imparted to a spin, Φ, is directly proportional to the position, x, the gyromagnetic ratio, $\frac{\gamma}{2\pi}$, the gradient-waveform function, $G_x(\delta)$, and the duration of the gradient, δ. For rectangular gradients of amplitude G_x and duration δ, this equation simplifies to

$$\Phi = \frac{\gamma}{2\pi} G_x x \delta. \qquad [2.4.2.2]$$

Imparting a phase to the spins is known as phase "tagging." Once a spin is "tagged", an identical gradient applied after a 180° RF pulse will serve to remove the phase "tag" (Fig 2.4.2.1). In the absence of translational motion, the net phase accumulation will be

$$\Delta\Phi = \frac{\gamma}{2\pi} G_x x_1 \delta - \frac{\gamma}{2\pi} G_x x_1 \delta = 0. \qquad [2.4.2.3]$$

SPIN MANIPULATION

On the other hand, if the molecule translates from its original position, x_1, to a new position, x_2, between the "tagging" and "untagging" gradient pulses, a net phase proportional to the displacement will be imparted to the spins according to

$$\Delta\Phi = \frac{\gamma}{2\pi}G_x x_2 \delta - \frac{\gamma}{2\pi}G_x x_1 \delta = \frac{\gamma}{2\pi}G_x (x_2 - x_1)\delta \qquad [2.4.2.4]$$

where $x_2 - x_1$ is the displacement of the spin, δ is the duration of the gradient, G_x is the gradient waveform function, and $\frac{\gamma}{2\pi}$ is the gyromagnetic ratio. This translational period between the spin tag and the removal of the tag is known as the diffusion time, Δ.

Since the overall signal is comprised of the individual spins, the measured magnetization can be written (Le Bihan and Basser, 1995) as the sum of the complex phases of the individual spins

$$\frac{M_{xy}}{M_0} = \sum_{n=1}^{N} e^{i(\Delta\Phi)_n} \, . \qquad [2.4.2.5]$$

Since diffusion is a random process, the distribution of displacements will be Gaussian, hence, the distribution of phases will also be Gaussian. From this we can write the normalized NMR signal as

$$\frac{M_{xy}}{M_0} = \int_{-\infty}^{\infty}\int_{-\infty}^{\infty} \left(e^{i\frac{\gamma}{2\pi}G(x_1-x_2)\delta} * \left[\frac{1}{\sqrt{4\pi D\Delta}} e^{-\frac{(x_1-x_2)^2}{4D\Delta}} \right] \right) dx_1 dx_2 \qquad [2.4.2.6]$$

where $e^{i\frac{\gamma}{2\pi}G(x_1-x_2)\delta}$ is the phase contribution of the displaced spins to the signal while $\frac{1}{\sqrt{4\pi D\Delta}} e^{-\frac{(x_1-x_2)^2}{4D\Delta}}$ represents the Gaussian behavior of the displacements.

Evaluating this function gives us

$$\frac{M_{xy}}{M_0} = e^{-(\gamma G\delta)^2 D\Delta} \qquad [2.4.2.7]$$

if we neglect the movement of spins while the gradient is being applied (Le Bihan and Basser, 1995). The finite duration of the gradient pulses can be compensated for by altering Eq. [2.4.2.7] to

$$\frac{M_{xy}}{M_0} = e^{-(\gamma G\delta)^2 D\left(\Delta-\frac{\delta}{3}\right)} \qquad [2.4.2.8]$$

for rectangular-shaped gradient pulses and

$$\frac{M_{xy}}{M_0} = e^{-\left(\frac{2}{\pi}\gamma G\delta\right)^2 D\left(\Delta-\frac{\delta}{4}\right)} \qquad [2.4.2.9]$$

for half-sine-shaped gradient pulses. The above equations can be re-written as

$$\frac{M_{xy}}{M_0} = e^{-bD} \qquad [2.4.2.10]$$

where b represents the gradient terms (i.e. $\left(\frac{2}{\pi}\gamma G\delta\right)^2\left(\Delta - \frac{\delta}{4}\right)$ for half-sine-shaped

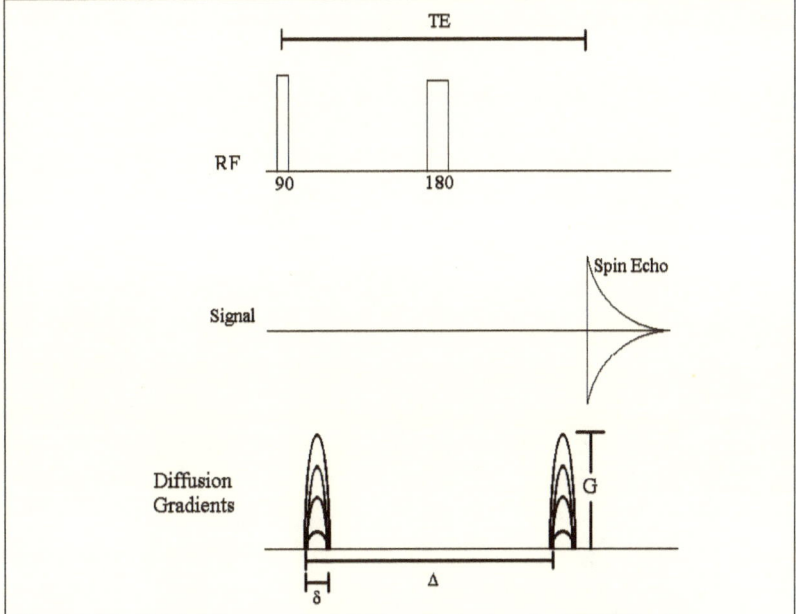

Fig. 2.4.2.1. Stejskal-Tanner pulsed-field gradient spin-echo (PFGSE) pulse sequence. The first set of gradients imparts a phase to the spins ("tag") while the second set of gradients removes the phase ("untag") of the spins. The net phase shift after the second gradient pulse reflects the net displacement. The attenuation of the echo is dependent on Δ, δ, G, and D.

gradients).

D can be measured by acquiring the signal at varying b-values and fitting to Eq. [2.4.2.10].

Chapter 3

Murine Tumor ^{19}F Experiments

The following experiment was started by Dr. Bernard Dardzinski during his stay at WPI as a Ph.D. student. Not only was Dr. Dardzinski responsible for much of the pioneering work leading up to this experiment, but he was also integrally involved in the design and execution of this particular experiment. One of the problems encountered during the course of this experiment was in the data analysis stage. Since both spectroscopic and imaging data were collected for the murine tumor oxygenation measurements, we expected that the information obtained from both methods would be comparable. My contribution to this project was in the analysis of the data and reconciling the differences between the spectroscopic and imaging data. Dr. Dardzinski was responsible for the acquisition of the data. While Dr. Dardzinski started the writing, I was responsible for analyzing the data and completing the writing of this manuscript.

3.1. Radiation Therapy

3.1.1. Abstract

Radiotherapy studies in mice using perfluoro-15-crown-5-ether showed a decrease in pO_2 following a single large dose of radiation. In conjunction with spectroscopic data, Inversion-Recovery Echo-Planar-Imaging data were collected at 1-3 hours, 10-13 hours, and 19-26 hours post irradiation, and T_1-maps generated in order to display localized changes in pO_2. The calculated T_1-maps were then weighted by their respective M_0-maps to find the weighted average of the T_1-maps, and an equivalent pO_2 of the tumor was then calculated from the weighted average. Untreated control animals that were subjected to the same time course showed no evidence of pO_2 decline, while the tumors irradiated with a single dose of 6 MeV electrons showed a decline in pO_2 by approximately 9 torr almost immediately after irradiation. The calculation of pO_2 using the weighted average of the T_1-maps was not only highly correlated to the spectroscopic measurements, it was approximately equivalent to the spectroscopic measurements. It is speculated that the decrease in the tissue oxygenation following radiation therapy is due to vascular damage caused by such a high dose of radiation, or edema within the interstitium of the tumor. Edema can cause the interstitial pressure to increase, resulting in vascular collapse. This in turn would lead to decreased perfusion and thus decreased oxygen delivery.

3.1.2. Introduction

Fluorine-19 MRI of exogenously administered PFC emulsions have seen wide-spread use in biomedical applications, especially since their introduction as artificial blood substitutes. By administering these agents intravascularly, ^{19}F MRI can provide anatomical or physiological information wherever these compounds perfuse or are sequestered. Imaging of the vascular system has been performed in large blood vessels in the heart and brain (Joseph, 1985a; Joseph et al., 1985b; Eidelberg et al., 1988a; Eidelberg et al., 1988b) and imaging of murine tumor vascularity to assess heterogeneity, necrosis, and the effects of photodynamic therapy has been demonstrated (Ceckler et al., 1990). ^{19}F MRI has also been used to image fluorine in lungs of mice submersed in oxygen-enriched PFCs (Thomas et al., 1986). Since PFC emulsions are foreign substances in the body of the host, large amounts of PFCs are sequestered in the reticuloendothelial system (RES) of the liver and spleen. This sequestration has made ^{19}F MRI amenable to studying the biodistribution of PFCs in these organs (McFarland et al., 1985; Ratner et al., 1987a; Sotak et al., 1993a; Sotak et al., 1993b; Barker et al., 1993; Dardzinski et al., 1993a). Macrophage activity present in tumors and abscesses has allowed the use of ^{19}F MRI to image PFCs sequestered in these lesions (Sotak et al., 1993b; Longmaid et al., 1985; Ratner et al., 1988; Sotak et al., 1991; Dardzinski et al., 1993b; Dardzinski et al., 1993c).

3.1.2.1. Oxygen Sensitive ^{19}F MRI and MRS

One of the unique applications of PFC emulsions in medicine and biology involves the use of these compounds to noninvasively measure oxygen tension *in vivo* using ^{19}F MRI/MRS. *In vivo* oxygenation measurements rely on the fact that the spin-lattice relaxation rate, R_1 ($1/T_1$), of vascular or tissue-sequestered PFC emulsions is linearly related to dissolved oxygen concentration (Parhami *et al.*, 1983; Clark *et al.*, 1984). This fundamental attribute has been utilized to perform oxygen-sensitive MRI/MRS of PFC emulsions in the mouse lung (Thomas *et al.*, 1986), vascular oxygen tension in rats (Fishman *et al.*, 1987), vascular oxygen tension in tumors (Fishman *et al.*, 1987), pO_2 imaging of cerebral blood oxygenation (Eidelberg *et al.*, 1988a; Eidelberg *et al.*, 1988b), partial oxygen pressure determination in the vitrectomized rabbit eye (Berkowitz *et al.*, 1991), and measurement of myocardial oxygen tension (Mason *et al.*, 1992). Sequestration of PFC emulsions has allowed oxygen-tension assessment in murine liver and spleen (Dardzinski *et al.*, 1993a; Clark *et al.*, 1985a; Holland *et al.*, 1993; Dardzinski *et al.*, 1994b; Barker *et al.*, 1994) and in murine tumors (Dardzinski *et al.*, 1993b; Dardzinski *et al.*, 1993c; Dardzinski *et al.*, 1994b; Mason *et al.*, 1991; Hees *et al.*, 1993; Dardzinski *et al.*, 1994a; Dardzinski *et al.*, 1994d).

While ^{19}F MRS in murine tumors is a powerful tool for measuring tumor oxygenation on a more global scale, being able to make pO_2 maps using MRI of these sequestered PFCs allows observation of localized changes in tumor oxygenation which are not possible with spectroscopy alone. Furthermore, the ability to acquire both imaging and spectroscopic data in a short time allows for a more complete understanding of tumor oxygenation and, in some sense, tumor physiology. In studies involving therapeutic interventions, the ability to assess changes in tumor oxygenation on a local scale enhances the data obtained using spectroscopy.

3.1.2.2. Hypoxia and Radioresistance

Based on previous works, there is empirical evidence that ionizing radiation in the presence of oxygen kills more cells than in the case of hypoxic cells. In a classic work, Gray found a correlation between radiosensitivity and tumor oxygenation (Gray *et al.*, 1953). A small increase in oxygen tension has been found to increase the radiosensitivity. Low oxygenation levels can increase the radioresistance up to three times as compared to normoxic tissue (Rubin and Casarett, 1968).

In the case of radiation therapy, the most important effect of radiation is to produce an activated form of oxygen. These activated oxygen species are commonly known as free radicals and have a free electron which promotes extremely efficient binding with cell components, such as strands of DNA or protein fragments. This binding of the free radicals to cell components can inhibit mitosis or the reparation of cell membranes. The ultimate point of no return in cell injury is the breakdown of cell membranes. Whatever the cellular mechanism, the most important predictor of the efficacy of radiation therapy is the availability of adequate amounts of molecular oxygen.

Tumor cells tend to be more hypoxic than normal tissue, thereby rendering them radioresistant. This hypoxic state can be linked to irregular vascularization of the tumor. The vascular supply in tumors tends to be very heterogeneous and cells tend to proliferate faster than the creation of new blood vessels, which supply nutrients to the cells. These cells are then located further away from vessels than the diffusion distance of oxygen. Oxygen will be consumed by intervening cells and will leave the furthest cells either necrotic or extremely hypoxic. This chronic or "diffusion"-limited hypoxia in tumor cords around capillaries was described by Thomlinson and Gray (1955). Their results indicated that cells furthest from the blood supply are poorly oxygenated. The diffusion distance of oxygen, in tubes of prepared radiation-induced fibrosarcoma (RIF-1) tumor tissue, was measured at 107 μm by fluorescent drugs which bind to hypoxic cells (Olive et al., 1992). Cells in close proximity to capillaries are well oxygenated. Cells that are past the diffusion limit of oxygen are hypoxic and those that are the furthest away from the vasculature ultimately perish and become necrotic.

Another mechanism of hypoxia has been proposed in which cells can become temporarily hypoxic by intermittent closing and opening of blood vessels. This second type of hypoxia depends on the structure of the vasculature itself and is identified as acute or "perfusion" limited hypoxia (Brown, 1979; Chaplin et al., 1987). This intermittent opening and closing leaves cells in close proximity to these vessels hypoxic but still in a viable state. If the outcomes in radiation therapy can be correlated to the oxygenation state of tumor cells, then the diagnosis of hypoxia could prove to be extremely useful to the radiation oncologist. Diagnosis of hypoxia with high spatial and temporal resolution would also allow the study of the effects of chemical adjuvants known to improve the state of acute hypoxia.

3.1.3. Experimental Methods

3.1.3.1. Hardware and Materials

All animal data were obtained on a GE-CSI-II 2.0T/45 cm imaging spectrometer with ±20 G/cm self-shielded gradients operating at 80.5 MHz for ^{19}F. The tumor studies were performed using a 5-turn, 18-mm diameter solenoid coil fitted with a copper shield. The animal was placed in a feedback-temperature-controlled airflow system within a Plexiglas® cylinder placed inside the usable 15-cm-magnet bore. Perfluoro-15-crown-5-ether (perfluoro-1, 4, 7, 10, 13- pentaoxacyclopentadecane, HemaGen/PFC, St. Louis, MO) was used as the ^{19}F NMR agent in the emulsified form.

3.1.3.2. Animal Preparation

A dose of 10g/kg of 40% (v/v) PFC emulsion was administered via tail vein injection for all *in vivo* studies. Conversion of percent oxygen tension to torr (mm Hg) assumed that

100% O_2 was equivalent to 760 torr. The magnetic field was shimmed on the tumor fluorine signal so that the corresponding resonance line width was between 20 – 30 Hz. Typical 90° RF pulse widths were approximately 15 microseconds at 100 W of power.

RIF-1 tumor cells (10^5 - 10^6 in 0.15 ml) were injected subcutaneously into lower backs of C_3H mice. RIF-1-tumor-bearing C_3H mice were injected intravenously with a 10 g/kg dose (0.35 ml for a 25 g mouse) of perfluor-15-crown-5-ether when the tumors had reached a volume of 0.5 to 1.5 cc. The PFC was allowed to clear from the vasculature for 3 to 7 days prior to imaging. Animals were anesthetized with 1.5% isofluorane, initially in air at a flow rate of 1.0 liters/min. Body temperature was maintained at 37°C (monitored via a rectal thermocouple probe) by circulating warm air at a temperature between 34 – 35°C.

3.1.3.3. Data Acquisition

A non-slice selective IR-EPI sequence (Stehling et al., 1990) was used to acquire the seven images used for calculating one pO_2 map. Coronal projection images were acquired using a saw-toothed scan of k-space. Seven IR-EPIs were collected with the following TI values: 0.08, 0.20, 0.50, 1.00, 2.00, 4.00, and 8.00 seconds. Other parameters were as follows: FOV = 30.0 mm with a pixel resolution of 64 x 64, TR = 10.0 s, TE = 70 ms, NEX = 8, SW = ±30 kHz. Due to the solenoid coil construction and intervening copper shield, the effective slice thickness was approximately 4 mm. After the acquisition of each of the seven IR-EPIs, a spectroscopic measurement of T_1 was performed using the same seven inversion times.

3.1.3.4. Radiation Therapy

Three to four sets of IR-EPI and IR spectroscopic measurements were performed as control measurements prior to irradiation. A special Lucite® restraining device was built to deliver the radiation therapy to the subcutaneous RIF-1 tumor located on the lower back of individual unanesthetized mice. The mice were confined in a prone position, and a lead shield was used to protect the head, thorax, and upper abdomen from the radiation beam. A linear accelerator (CGR Saturne 1, 6 MeV electrons) was used to irradiate the mice with a calculated single dose of 1000 cGy to be delivered to a volume described by a 90% isodose line which encompassed the whole tumor volume with only 40% of the dose reaching the gut and hematopoietic tissue. A 5-mm-thick piece of bolus (tissue equivalent) material was placed on top of the tumor so the depth at maximum dose, D_{max}, would be at the center of the tumor. D_{max} for this accelerator was calculated to be 14 mm. The tumor oxygenation was then mapped and measured spectroscopically for 1 to 3 hours post-irradiation at 15-minute time intervals. Animals were then removed from the magnet and allowed free access to food and water. At approximately the 12 and 24 hour post-irradiation the NMR procedure was repeated with approximately four measurements of pO_2 being taken over one hour at each time point. The same exact procedure was followed for the control animals except that no radiation therapy was administered.

3.1.3.5. Data Analysis

Data were analyzed spectroscopically, to determine the global effects of the radiation therapy, and using imaging to determine the local effects of radiation therapy on the tumor. The weighted average of the imaging data was then calculated to confirm that the imaging results and the spectroscopic results were in good agreement (Han *et al.*, 1997).

For the calculation of the T_1-map and the weighted average of the T_1-map, a three-parameter curve fitting routine was used to calculate the PFC T_1 values, on a pixel-by-pixel basis, using the equation:

$$M_{Zi}(t) = M_{0i}(1 - (B^*\exp(-t/T_{1i}))) \qquad [3.1]$$

The PFC T_1 value from each pixel was then weighted by its respective spin density and a weighted average of the T_1 value was calculated as:

$$T_{1\ \text{Weighted Average}} = (\Sigma(T_{1i}*M_{0i}))/(\Sigma M_{0i}) \qquad [3.2]$$

From the T_1 values measured via spectroscopy and imaging, the pO_2 was calculated from the calibration equation:

$$pO_2 = (297.38 / T_1) + (2.9398 * \textbf{Temperature}) - 211.35 \qquad [3.3]$$

Because the absolute pO_2 values differ from tumor to tumor, the baseline pO_2 measurements prior to irradiation were averaged together for each animal individually. The pO_2 values determined post-irradiation were then subtracted from this average baseline value. This produced a ΔpO_2 measurement that was averaged over each group for each individual time point post-irradiation. All data are presented as the mean ± the standard error of the mean (SEM).

3.1.4. Results

RIF-1 tumors that were irradiated with a single large, 1000 cGy, dose of 6 MeV electrons exhibited an overall decline in tissue oxygenation when compared to pre-irradiated baseline oxygen tension values. Animals that were subjected to the same time course and

Table 3.1. Spectroscopically-determined change in tumor oxygen tension (ΔpO_2) from baseline values measured at the indicated time periods post-irradiation (1000 cGy of 6 MeV electrons) for the four animals given radiation therapy (Rad1, Rad2, Rad3, and Rad4) and two control animals (Cont1, Cont2).

*   Rad2 died 11 at hours post-irradiation
**   Cont1 died at 14 hours post-simulated-irradiation

	1-3 hours post-irradiation	10-13 hours post-irradiation	19-26 hours post irradiation
Rad1	-16.8 torr	-10.7 torr	-20.5 torr
Rad2	-1.4 torr	*	
Rad3	-9.8 torr	-8.1 torr	-7.1 torr
Rad4	-10.1 torr	-6.7 torr	-11.1 torr
Cont1	-2.3 torr	-6.2 torr	**
Cont2	-0.8 torr	-2.6 torr	2.9 torr

Table 3.2. Changes in tumor oxygen tension (ΔpO_2) from baseline values measured at the indicated time periods post-irradiation (1000 cGy of 6 MeV electrons) for the four animals given radiation therapy (Rad1, Rad2, Rad3, and Rad4) and two control animals (Cont1, Cont2). Reported values are the ΔpO_2 measured using the weighted average of the T_1-maps.

*   Rad2 died 11 at hours post-irradiation
**   Cont1 died at 14 hours post-simulated-irradiation

	1-3 hours post-irradiation	10-13 hours post-irradiation	19-26 hours post irradiation
Rad1	-21.5 torr	-14.3 torr	-7.4 torr
Rad2	-0.8 torr	*	
Rad3	-5.2 torr	-1.1 torr	-2.5 torr
Rad4	-3.4 torr	1.9 torr	-3.7 torr
Cont1	-0.4 torr	-2.8 torr	**
Cont2	2.6 torr	1.2 torr	5.3 torr

Table 3.3. Summary of tumor pO_2 changes (ΔpO_2) measured spectroscopically and using the weighted average of the T_1-maps. All animals treated with radiation therapy were subjected to 1000 cGy dose of 6 MeV electrons. The animals in the control group underwent simulated radiation therapy.

*   No SEM associated with this measurement due to the fact that n = 1.

SPECTROSCOPY		
Time Post Irradiation	Radiation Therapy (n=4)	Control (n=2)
1-3 hours	-9.5 ± 3.2 torr	-1.5 ± 0.8 torr
10-13 hours	-8.5 ± 1.2 torr	-4.4 ± 1.8 torr
19-26 hours	-12.9 ± 4.0 torr	2.9 torr*
WEIGHTED AVERAGE		
Time Post Irradiation	Radiation Therapy	Control
1-3 hours	-7.7 ± 4.7 torr	1.1 ± 1.5 torr
10-13 hours	-4.5 ± 5.0 torr	-0.8 ± 2.0 torr
19-26 hours	-6.4 ± 3.3 torr	5.3 torr*

physical restraint showed very little evidence of decline. Table 3.1 indicates the change in oxygen, ΔpO_2, for specific time intervals for the four irradiated animals and the two control animals while Table 3.2 shows the corresponding results from the weighted average of the imaging maps. The decrease in tumor oxygen tension for all spectroscopically-measured radiation therapy animals ranged from -9.5 ± 3.1 torr, at 1-3 hours post irradiation, to -8.5 ± 1.2 torr, at 10-15 hours post-irradiation, and finally decreased to -12.9 ± 4.0 torr, at 19-26 hours post irradiation. The spectroscopic measurements for the controls at 1-3 hours post-simulated-radiation therapy were -1.5 ± 0.8 torr, at 10-15 hours post irradiation, -4.4 ± 1.8 torr, and for the one animal that survived, the ΔpO_2 actually increased to 2.9 torr compared to the baseline value. The

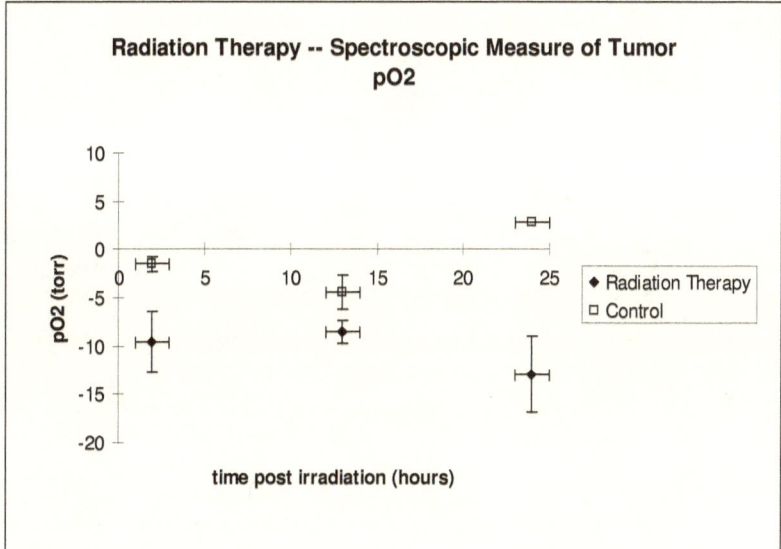

Fig. 3.1. Plot of spectroscopically-determined ΔpO_2 as a function of time post-irradiation with a single large dose (1000 cGy) of 6 MeV electrons. Animals that were irradiated showed a marked decrease in tumor pO_2 while the control animals showed very little change from the baseline.

corresponding values for the weighted average of the ΔpO_2 values from the imaging maps were -7.7 ± 4.7, -4.5 ± 5.0, and -6.4 ± 3.3 torr at 1-3, 10-13, and 19-26 hours post irradiation, respectively, for the radiation therapy animals. The corresponding ΔpO_2 values for the control animals were 1.1 ± 1.5, -0.8 ± 2.0, and 5.3 torr for the same time periods. The results of the difference of the weighted average of the pO_2 maps gave similar results to that of the spectroscopic measurements (Table 3.3, Figs. 3.1 and 3.2).

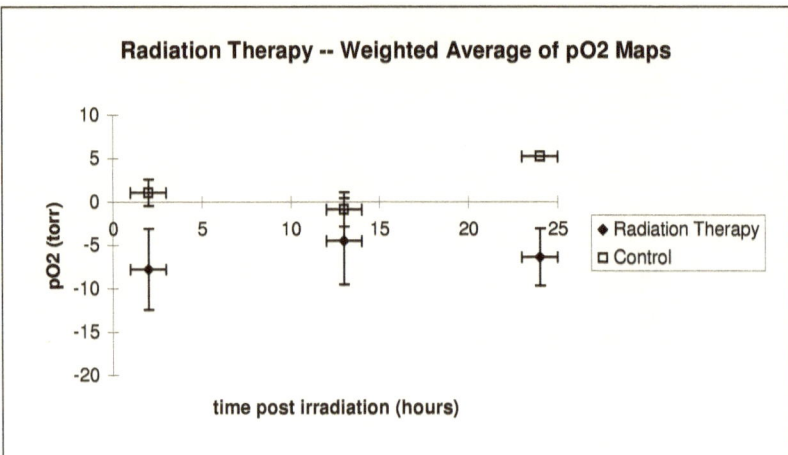

Fig. 3.2. Plot of ΔpO_2 determined using the weighted-average of the T_1-maps as a function of time post irradiation with a single large dose (1000 cGy of 6 MeV electrons). Radiation-therapy animals showed a decrease in tumor oxygenation following radiation therapy, while the controls showed a small fluctuation around ΔpO_2 of zero torr. The error bars are larger for the imaging data (Fig. 3.1) due to the fact that the S/N is smaller for the imaging compared to that of the spectroscopy.

Fig. 3.3 depicts the spatial changes in pO_2 for one of the animals from pre-irradiation to the 24-hour time point post-irradiation. Although the repositioning of the animal was not perfect at the different time points of the measurement, it is clear that the tumor which was better oxygenated before radiation therapy (Rad 1) showed a marked decrease in tumor oxygenation while the control (Cont 2) and the initially hypoxic tumor (Rad 2)

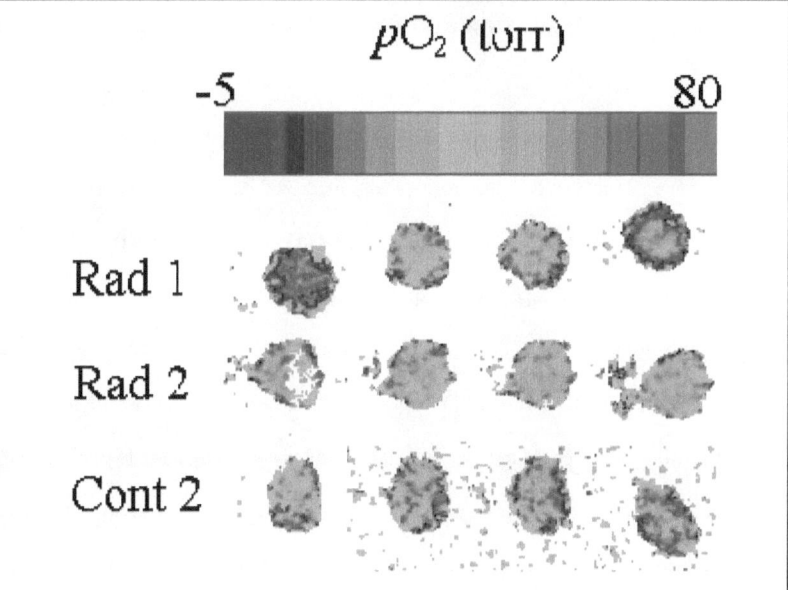

Fig. 3.3. pO_2 maps of typical tumors during the time course of the radiation therapy study. Rad 1: Tumor with initially normal oxygen tension. pO_2 decreases over the whole tumor with the initial dose of radiation. Also, it is interesting to note that the areas of the tumor which were initially well oxygenated showed a larger response to radiation therapy than the regions which were initially more hypoxic. Rad 2: Tumor with initially low oxygenation. pO_2 is relatively unaffected by radiotherapy. Cont 2: Tumor that was not irradiated. Tumor shows no significant changes in pO_2 over the time course of 24 hours.

showed no significant change in oxygenation. Another point to note is that the portions of the tumor (Rad 1) which were initially well oxygenated showed a larger response to radiation therapy compared to the areas which had lower initial oxygen tension.

3.1.5. Discussion and Conclusion

Figs. 3.1 and 3.2 indicate that immediately after radiation therapy, the oxygen tension decreases dramatically. Spectroscopically-measured oxygen tension declined an average of 12.2 torr in three of the four animals (Rad1, Rad3, and Rad4) at 1-3 hours post-irradiation. The average baseline spectroscopic pO_2 value before irradiation was 24.2 torr for this group while the average baseline spectroscopic pO_2 value for Rad2 was only 2.0 torr. The average spectroscopic ΔpO_2 for Rad2 was only -1.4 torr at 1-3 hours post-irradiation. This could indicate that since Rad2 was initially more hypoxic than the other three tumors, there was less radiation damage. Furthermore, the areas of the tumors that were initially well oxygenated showed a greater response to radiation therapy, which indicates that the oxygen content and effects of radiation therapy are highly correlated. It is well known that cells that are highly oxygenated have a three-fold increase in radiosensitivity over hypoxic cells. The average spectroscopic ΔpO_2 value of the control groups (-1.0 ± 1.2, n = 38 replicate measures) was statistically different from the treated group (-8.4 ± 0.9, n = 72 replicate measures, P < 0.00001, ANOVA). Also, for the weighted-average of the T_1-maps, a statistical difference (mean$_{control}$ = 0.9 ± 0.8 torr, n=37; mean$_{irradiation}$ = -5.6 ± 1.0 torr, n = 72; P < 0.00005, ANOVA).

Since there is such a dramatic decrease in tumor oxygenation for RIF-1 tumors irradiated with a 1000 cGy (10 Gy) dose of 6 MeV electrons, the mechanism of decline is not fully understood. Rubin and Casarett have shown that there is vascular damage after large single doses of radiation (15 Gy) in rat Walker carcinoma and Murphy-Sturm lymphosarcoma tumors (Rubin *et al.*, 1966). This can lead to lower tumor oxygen tension levels and would effectively lower the radiosensitivity of future treatments implying the need to monitor oxygen levels and treatment with lower fractional doses. Perhaps a large single dose of radiation causes edema to form within the interstitial space of the tumor tissue (Rubin *et al.*, 1968), leading to fibrosis. Increasing vascular permeability to albumin in rabbit skeletal muscle following single large doses of 2-30 Gy has been measured isotopically following irradiation (Krishnan *et al.*, 1987). Their results indicated the amount of extravasation of albumin is significant above 6 Gy, and changes in vascular permeability occurs immediately after irradiation with doses as small as 2 Gy.

Increased extravasation of plasma proteins can lead to interstitial edema, especially in tumors, where there is little or no lymphatic clearance. Edema formation can increase interstitial fluid pressure (Jain, 1987; Boucher *et al.*, 1990) and vascular collapse (Boucher *et al.*, 1992), resulting in decreased perfusion, and therefore, decreased oxygen delivery. The immediate decrease in RIF-1 tumor oxygenation as demonstrated in Fig.

3.1 and Fig. 3.2 could be the result of vascular damage or increased edema formation. Later, with decreasing edema, the tumor oxygenation begins to recover at the periphery faster than in the center of the tumor. Tumor physiology indicates that most of the large vascular network is located in the periphery of the tumor. Oxygenation in this area should tend to recover first because of the large number of high caliber vessels that are located in the region, and since the interstitial fluid pressure is lower in the periphery of solid neoplasms (Boucher *et al.*, 1990). At approximately 10-15 hour post-irradiation, tumor oxygenation seems to recover slightly but never reaches the pre-irradiation oxygen tension values. At 19-26 hours post-irradiation, the oxygen tension tends to decline again. However, other mechanisms as suggested by Kallman (1972) could lead to increased oxygenation. These include the increase in the diffusion limit of oxygen past the now lethally injured cells, improved circulation and hence more oxygen carrying capability of the capillaries, shrinkage of cells accompanied by a shift in capillaries and oxygenation centripetally, and finally, migration of surviving cells into previously hypoxic zones providing improved oxygenation.

The objective of these experiments was to ascertain the time of optimal tumor oxygenation following radiotherapy for planning fractionated treatment. However, most fractionated therapy strategies use doses that are 10 to 25% smaller than the dose given in this set of experiments. When multiple doses are given, the total dose can reach 50-60 Gy, but each single dose will probably not disrupt the vascular network. Continued experiments with smaller doses of radiation should be performed to determine if changes in tumor pO_2 can be detected following irradiation, in the absence of edema, using ^{19}F NMR pO_2 mapping techniques.

In summary, IR-EPI of perfluoro-15-crown-5-ether provides a rapid method for producing oxygen-tension maps *in vivo*. Dynamic changes in the oxygenation state of tumors can be studied with a temporal resolution of 10 minutes. This allows for dynamic monitoring of the tumor oxygenation state following therapeutic interventions. Spatial changes in oxygenation can also be visualized. Non-invasive oxygen-tension methods of this type should help maximize the response to radiation therapy, since the levels of hypoxia can be monitored. This technique should also prove useful in a clinical setting to ultimately correlate the effects of hypoxia on radiation-therapy outcomes.

3.1.6. Acknowledgements

The authors would like to thank J. Murray, J. Kelley, and A. Isabelle for their help in performing the radiation therapy, and R. J. Kaufman, Ph.D. and HemaGen (St. Louis, MO) for providing the perfluoro-15-crown-5-ether used in this work. Funding for this research by The Whitaker Foundation is gratefully acknowledged.

Chapter 4

Murine Tumor Diffusion and pO_2 Experiments

The following is a pre-print of the article accepted for publication in *NMR in Biomedicine*. The experiment deals with the correlation between the water diffusion coefficient and oxygen tension. I was responsible for the calibration curves, most of the animal care, and data acquisition. These responsibilities included inoculating the C_3H-mice with RIF-1 tumors and also administering the perfluorocarbon emulsions *via* tail vein injections. I also contributed to the data analysis by processing the weighted-average of the pO_2 maps. The manuscript was written by Drs. Karl Helmer and Christopher Sotak.

4.1. On the Correlation Between the Water Diffusion Coefficient and Oxygen Tension in RIF-1 Tumors

Karl G. Helmer[1], Sam S. Han[1], Christopher H. Sotak[1,2]

[1] Department of Biomedical Engineering, Worcester Polytechnic Institute, Worcester, MA 01609
[2] Department of Radiology, University of Massachusetts Medical School, Worcester, MA 01605

Running Title: On the correlation between the ADC and pO_2 in RIF-1 tumors

Address correspondence to:
    Karl G. Helmer, Ph.D.
    Department of Biomedical Engineering
    Worcester Polytechnic Institute
    100 Institute Road
    Worcester, MA 01609
    Tel: 508 831 5716
    Fax: 508 831 5541
    email: kgh@wpi.edu

Keywords: oxygen tension mapping, ^{19}F, water diffusion coefficient, RIF-1 tumor

4.1.1. Abstract

Water diffusion-coefficient mapping was used in conjunction with ^{19}F inversion-recovery echo-planar imaging (IR-EPI) of a sequestered perfluorocarbon (PFC) emulsion to investigate the spatial correlation between the diffusion coefficient of water and the tissue oxygen tension (pO_2) in radiation-induced fibrosarcoma (RIF-1) tumors (n = 11). The diffusion-time-dependent apparent diffusion coefficient, $D(t)$, was determined by acquiring diffusion coefficient maps at 20 different diffusion times. Maps at four representative time points in different regions of the $D(t)$ curve were selected for final analysis. An intravenously administered PFC emulsion, perfluoro-15-crown-5-ether, was used to generate the pO_2 maps. $D(t)$ and pO_2 data were acquired with the animal breathing either air or carbogen (95% O_2 – 5% CO_2) to investigate the effects of increased tumor pO_2 on $D(t)$. The average increase in tumor pO_2 was 22 torr when the breathing gas was changed from air to carbogen. Correlating plots generated from pixel data for $D(t)$(air breathing) versus $D(t)$(carbogen breathing) showed little deviation from a slope of unity. Correlation plots of $D(t)$ versus pO_2 indicate that no correlation is present between these two parameters. This study also confirms that necrotic tissue was best differentiated from viable tumor tissue based on $D(t)$ maps at long diffusion times.

4.1.2. Introduction

The assessment of tissue oxygen tension (pO_2) is an important component in the determination of radio- and chemotherapeutic efficacy (Vaupel, 1977; Sostman *et al.*, 1991). The experimental determination of pO_2, however, is a difficult and often invasive procedure involving either electrodes or implanted EPR probes (Sostman *et al.*, 1991; Terris *et al.*, 1992; Bacic *et al.*, 1993) or exogenously administered compounds (Mason *et al.*, 1991; Baldwin and Ng, 1992; Dardzinski and Sotak, 1994; Hees and Sotak, 1993). It would therefore be advantageous to have a noninvasive and more easily measured indicator of the oxygen distribution in tumor tissue. To this end, Dunn *et al.* (1995) recently showed that the apparent diffusion coefficient (*ADC*) of water in chronically hypoxic tissue is directly related to tumor pO_2. The existence of a relationship between water *ADC* and tumor oxygenation would be valuable in differentiating the oxygen status of viable, hypoxic, and necrotic tissue as well as monitoring therapy.

In the initial study by Dunn *et al.*, calculated *ADC* maps were produced for each of seven RIF-1 tumors at a single diffusion time (15 ms). Oxygen tension measurements were obtained at two locations within each tumor (using EPR of implanted lithium phtalocyanine crystals), corresponding to the positions of the highest and lowest values in the calculated *ADC* map. The correlation coefficient between *ADC* and pO_2 showed a positive trend, i.e., large values of *ADC* correspond to large values of pO_2. However, the authors noted that such a correlation is restricted to areas where the tumor tissue was chronically hypoxic, but where there was no significant necrosis. The authors hypothesize that such an environment gives rise to impaired osmotic regulation in these cells with ensuing cellular swelling and a concomitant reduction in *ADC*. The basis for this hypothesis is similar to that for brain tissue, where cytotoxic edema is thought to be responsible for the decline in water *ADC* following an acute ischemic insult (Moseley *et al.*, 1990; Knight *et al.*, 1991).

In order to fully assess the potential of this method, we have investigated the relationship between water *ADC* and pO_2 under a wider range of experimental conditions than was employed in the above study. Calculated *ADC* maps from RIF-1 tumors were compared, on a pixel-by-pixel basis, with tumor pO_2 maps that were obtained from the same location using ^{19}F inversion-recovery echo-planar imaging (IR-EPI) of a sequestered perfluorocarbon (PFC) emulsion. This approach ensures that the relationship between *ADC* and pO_2 can be investigated for the full spectrum of viable, hypoxic, and necrotic tumor tissue and will allow us to characterize any limitations that are associated with this method.

The investigation of this relationship between *ADC* and pO_2 must also take into account the dependence of the water *ADC* on diffusion time in tumor tissue. In the present work, $D(t)$ is used to denote the *ADC* measured at a specific diffusion time, *t*, while *ADC* is used to denote the apparent diffusion coefficient without regard to the diffusion time. In all cases the *ADC* was measured by varying the applied field gradient only. A recent study using RIF-1 tumors (Helmer et al., 1995) has established that the behavior of $D(t)$ as t is changed is dependent upon tumor tissue type. For example, $D(t)$ values for necrotic tumor tissue are generally large and show little change with diffusion time, whereas $D(t)$ values for viable and hypoxic tissue can vary considerably with diffusion time and are generally lower than those for necrotic tissue. Given the potentially confounding effects of time-dependent diffusion on pO_2 measurements derived using this approach, the effect of the time-dependence of *ADC* on the correlation between tumor pO_2 and *ADC* was also investigated.

Finally, in order to relate changes in *ADC* values with changes in tumor oxygenation, time dependent *ADC* maps and pO_2 maps were compared for animals breathing either air or carbogen (95% O_2 – 5% CO_2). Carbogen breathing is known to increase the radiosensitivity of hypoxic cells in murine tumors (Suit et al., 1972; Siemann et al., 1977) by increasing respiration and cardiac output and, therefore, oxygen delivery (Kruv et al., 1967). The pO_2 mapping technique used in these studies has been shown to be sensitive to changes in tumor pO_2 following carbogen breathing (Dardzinski and Sotak, 1994) and hence provides a basis for identifying regions of the tumor where corresponding changes in *ADC* might also be expected.

4.1.3. Background

NMR diffusion measurements in fluid-filled porous media can provide useful structural information about the sample. The diffusion coefficient of the fluid in the interstitial space varies as a function of the diffusion time because of the interaction of the diffusing molecules with restricting boundaries at the medium. At short diffusion times, only molecules at the boundary surfaces are restricted and the value of $D(t)$, the time-dependent apparent diffusion coefficient, is reduced from D_0 (the bulk diffusion coefficient of the fluid) in direct proportion to the volume of the surface layer of restricted molecules. In this regime, the slope of a plot of $D(t)$ versus $t^{\frac{1}{2}}$ is proportional to the ratio of the surface area to pore volume, S/V (Mitra et al., 1992; Mitra et al, 1993), a "local" property of the medium $\left(\frac{D(t)}{D_0} \propto \left(\frac{S}{V} \right) \sqrt{D_0 t} \right)$. At long diffusion times, $D(t)$ reaches a constant, diffusion-time-independent value, D_{eff}, where each molecule has effectively experienced an equivalent portion of the confining medium. In this case, D_{eff} is reduced from D_0 in proportion to the tortuosity (Johnson et al., 1982; Haus and Kehr, 1987; Nicholson et al., 1979; Nicholson and Phillips, 1981; Nicholson and Rice, 1991), τ, (i.e., $D_{eff}=D_0/\tau$), of the connective pathways between pore spaces. Earlier work (Helmer et al., 1995) has found that using long diffusion times, such that the diffusing water molecules are in the tortuosity regime, is useful for differentiating necrotic from viable tumor tissue. This is the case since the measured ADC is reflecting the effects of restriction on a "global" rather than "local" scale.

Fluorine-19 NMR spectroscopy and imaging of PFC emulsions have been used extensively to measure tissue oxygenation (Mason *et al.*, 1991; Baldwin and Ng, 1992; Dardzinski and Sotak, 1994; Hees and Sotak, 1993). The spin-lattice relaxation rate R_1 ($1/T_1$), of PFCs is linearly related to the dissolved oxygen concentration (Parhami and Fung, 1983; Kong *et al.*, 1984) and thus allow oxygenation measurements to be performed in any tissue or organ in which the PFC is sequestered. In the case of solid murine tumors, fenestration in the tumor vasculature allow the intravenously administered PFC emulsion particles to leak into the extravascular space and serve as a noninvasive oxygen probe. Although a number of different PFC molecules have been used for measuring tumor oxygenation (Mason *et al.*, 1991; Baldwin and Ng, 1992; Hees and Sotak, 1993), more recent candidates, such as perfluoro-15-crown-5-ether (Dardzinski and Sotak, 1994), mitigate most of the problems associated with ^{19}F MRI of PFCs. In particular, perfluoro-15-crown-5-ether has a relatively long T_2-relaxation time compared with previously used PFCs and, since it contains 20 identical fluorine atoms, it has a single resonance, which eliminates chemical shift artifacts and J-modulation effects. ^{19}F EPI of perfluoro-15-crown-5-ether permits rapid and sensitive measurements of tumor pO_2.

4.1.4. Methods

RIF-1 tumor cells were prepared according to Twentyman *et al.* (1980) Cells were injected subcutaneously in the lower back of C_3H mice (n = 11) and allowed to grow to varying volumes (0.4 cc to 1.2 cc). Tumor volumes were determined by using the relation

$$V = \frac{\pi}{6}(a \cdot b \cdot c) \qquad [4.1]$$

where a, b, and c are the tumor length, width, and height, respectively. When the tumor had reached the desired volume, the tumor-bearing mice were administered a 15g/kg dose of a 40% (v/v) emulsion of perfluoro-15-crown-5-ether (perfluoro-1, 4, 7, 10, 13-pentaoxacyclopentadecane) (HemaGen/PFC, St. Louis, MO) *via* tail vein injection. Imaging experiments were performed three to seven days following PFC injection to ensure clearance from the vasculature. Animals were anesthetized during imaging with 1.5% isofluorane delivered in air at 1.0 L/min. Circulating air at 34°C was used to maintain the animal's body temperature at 37°C.

MRI data was acquired using a horizontal bore GE CSI-II 2.0T/45 cm imaging spectrometer (GE NMR Instruments, Fremont, CA) operating at 80.5 MHz for ^{19}F and 85.5 MHz for ^{1}H and equipped with ±20 G/cm self-shielded gradients. A four-turn, 15-mm-diameter solenoid coil was used for all experiments. Maps of the apparent diffusion coefficient were generated for twenty different diffusion times (from 11.0 ms to 560.5 ms) to delineate the $D(t)$ curve. The data from four representative diffusion-time points are analyzed in this paper. Twenty diffusion-weighted images were acquired for each map. $D(t)$ was extracted from the initial linear slope using a linear regression fit to the equation

$$\ln(M) = \ln(M_0) - k^2(D(t))t \qquad [4.2]$$

where k=$2\gamma g \delta/\pi$, M is the measured signal intensity and M_0 is the signal intensity without the applied diffusion gradient (see Helmer et al., 1995 for more details). The factor of $2/\pi$ in the expression for k takes into account the use of half-sine-shaped diffusion-sensitizing gradient pulses. Each image was obtained using either a Stejskal-Tanner sequence (Stejskal and Tanner, 1965) (t_{diff} = 11.0 – 57.0 ms) or a stimulated-echo variant (t_{diff} =87.5 – 560.5 ms), both employing EPI with a saw-tooth data acquisition scheme (Turner and Le Bihan, 1990). Echo times were the same (100 ms) for both sequences to ensure equal T_2-relaxation weighting. Diffusion gradients were incremented successively in 0.6 G/cm steps from 0.6 G/cm to 12.0 G/cm for diffusion times less than 100.0 ms. In order to keep the amount of attenuation constant, the initial and incremental gradient values were decreased for diffusion times greater than 100.0 ms. The gradient pulse width, δ, was 10.0 ms. Coronal EPIs were acquired with FOV = 30 x 30 mm^2, slice thickness = 3.0 mm, TR = 2.0 s, NEX = 2 (spin echo) or 4 (stimulated echo), and TE = 100.0 ms. The EPI data acquisition time was 65.5 ms, the spectral width was ±30 kHz, and the digital resolution was 64 x 64 data points. Images were acquired such that the center of the imaging slice coincided with the center of the tumor. Hematoxylin and eosin (H & E) staining of the tumor was performed to identify necrotic regions. Several histological slices were taken from within the imaging slice to check for local differences in necrotic and viable tissue volumes.

In vitro standard curves of R_1 (=$1/T_1$) vs. %O_2 for the neat perfluoro-15-crown-5-ether were obtained for four different standard gases, 0, 5, 21, and 30% O_2 (the balance being N_2) and four temperatures, 27, 32, 37, and 42°C. The gas bubbled into the PFC for 30 min at the require %O_2 and a spectroscopic measurement of T_1 was made at each of the above four temperatures. Multiple-linear regression was then performed on the data to extract the equation for pO_2 as a function of R_1 and temperature (T).

To generate R_1 maps from the RIF-1 tumors, ^{19}F images of the sequestered PFC were acquired using slice-selective IR-EPI. Imaging parameters include FOV = 30 x 30 mm^2, slice thickness = 3.0 mm, pre-delay = 10.0 s, acquisition bandwidth of ±70 kHz, EPI data acquisition time of 28.6 ms, TE = 70 ms, NEX = 8, pixel resolution of 64 x 64, and seven inversion times of 0.08, 0.20, 0.50, 1.00, 2.00, 4.00, and 8.00 s. The same inversion times and sequence parameters were used for both the calibration and *in vivo* experiments. Note that the same slice thickness and slice position was used for both the diffusion and R_1 maps.

R_1 maps were calculated, on a pixel-by-pixel basis, from the ^{19}F IR-EPIs using a Levenberg-Marquardt nonlinear least-squares fitting method (Press *et al.*, 1988). Pixel intensity, *S(TI)*, was fitted to the equation

$$S(TI) = A\left(1 + Be^{-(TI \cdot R1)}\right) \qquad [4.3]$$

where TI is the inversion time, and A and B are fitting constants. Each R_1 map was filtered by: 1) using a diffusion map of the same tumor as a mask in order to fit only those pixels originating from the tumor itself, and 2) excluding pixels in which there was no measured signal in either the air or carbogen data (corresponding to no sequestered PFC). An oxygen tension map was then calculated from the R_1 map using the *in vitro* calibration curves. Histograms of frequency versus pO_2 were generated by separating the pixel data into bins of 5 torr to display the range of values and to highlight the difference in tumor oxygenation due to the change in breathing gas.

Of importance in these experiments is the difference in pO_2 measured before and after a change in breathing gas. This difference was characterized using three different measures, each using all (non-zero) pixels in a given pO_2 map: the mean pO_2, the median pO_2, and the weighted-mean pO_2. Both the mean and median were calculated since the histograms of pO_2 frequency were not always normal distributions and the entire distribution was not affected equally by the change in breathing gas. The weighted-average of each map was constructed by weighting each pixel pO_2 by its spin density, M_0, and calculating the mean overall pixels, i, using

$$pO_2(\text{weighted-average}) = \frac{\sum (M_{0i} \cdot pO_{2i})}{\sum M_{0i}} \qquad [4.4]$$

Data were acquired first with the animal breathing air, using diffusion-weighted images to generate diffusion maps for each diffusion time. This was followed by the acquisition of the seven IR-EPI ^{19}F images used in the calculation of the pO_2 map. The breathing gas was then changed to carbogen, and the diffusion and ^{19}F data were again acquired in the same order. The start of data acquisition was approximately 10 min after the change in breathing gas.

4.1.5. Results

Multiple linear regression was used to extract the relationship between dissolved oxygen concentration and R_1 and T for four different temperatures and oxygen concentrations from three different trials. The resulting equation was

$$R_1 = 0.711 + 0.026 \cdot O_2 - 0.010 \cdot T \qquad r^2 = 0.998 \qquad [4.5]$$

where O_2 is in percent and T is in degrees Celsius. Solving Eq. [4.5] for pO_2,

$$pO_2 = 297.4 R_1 + 2.940 T - 211.4. \qquad [4.6]$$

Eq. [4.6] was used on a pixel-by-pixel basis to transform the R_1 maps into pO_2 maps. A temperature of 37°C was assumed in Eq. [4.6].

The changes seen in the computed pO_2 maps, when air is replace by carbogen as the breathing gas, are presented in Fig. 4.1 for a representative RIF-1 tumor. Fig. 4.1a is the map for the air-breathing mouse while Fig. 4.1c is the map for the same mouse breathing carbogen. The color scale beside the pO_2 map for the carbogen breathing mouse ranges from –20 to 80 torr and is the same for both maps. Note that the majority of the increase in pO_2 is evident in the periphery of the tumor where the vascular volume is greater

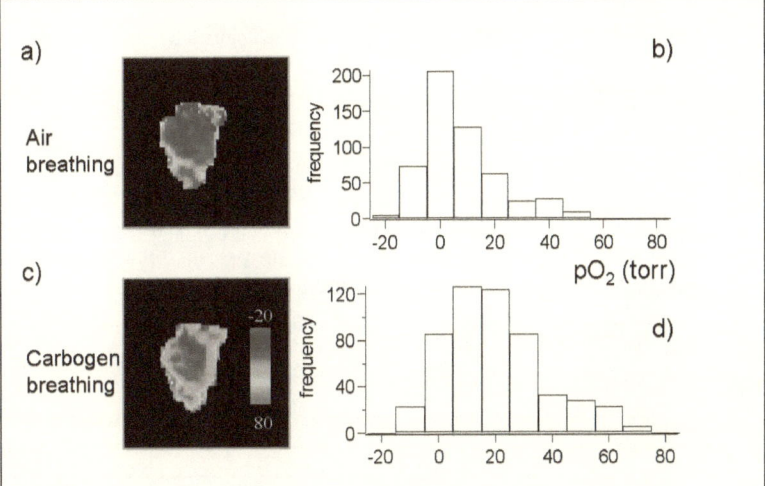

Fig. 4.1. Examples of oxygen tension maps in a RIF-1 tumor as a function of breathing gas. The color scale has a range of –20 to 80 torr. Only pixels that contain sequestered PFC were used to create the map. The slice thickness (3.0 mm) and position are the same as the slice used for the *ADC* maps. (a) Calculated pO_2 map acquired during air breathing. (b) Histogram of pO_2 values taken from the map in (a). Histogram bins are 10 torr wide. (c) Calculated pO_2 map acquired during carbogen breathing. (d) Histogram of pO_2 values taken from the map in (c). The mode of the peak has shifted from the bin centered around zero torr to the bin centered around 10 torr.

(Bhujwalla *et al.*, 1996). The pO_2 values were grouped into 5-torr bins and displayed as histograms in Fig. 4.1b (air breathing) and Fig. 4.1d (carbogen breathing). Due to their asymmetric distribution, the median rather than the mean is used as an index for the histograms. Table 4.1 lists the change in tumor pO_2 for each animal using the weighted-average, the unweighted-mean, and the median. For the 11 tumors studied here, the average increase in median pO_2 value when breathing gas was changed from air to carbogen was 20 ± 3 torr (mean ± SEM) with a p-value of 0.0001. Fig. 4.2 shows a histological slice from the tumor of Fig. 4.1. The colors have been reversed to provide the greatest contrast and hence light areas are regions of viable tissue.

Table 4.1. Changes in tumor pO_2 with a change in breathing gas from air to carbogen for 11 RIF-1 tumors. Numbers are calculated directly from the pixel pO_2 values. Weighted averages were calculated using Eq. 4.4.

Animal Number	Differences in Weighted Average (torr)	Differences in Unweighted Means (torr)	Differences in Medians (torr)
1	11	14	13
2	37	37	37
3	23	26	20
4	27	28	26
5	11	13	14
6	15	16	15
7	13	14	10
8	6	9	2
9	34	36	37
10	17	21	20
11	30	28	28
Mean (SEM)	20 (3)	22 (3)	20 (3)

Table 4.2. Fitting parameters for correlation plots of $D(t)$ for air breathing versus $D(t)$ for carbogen breathing for RIF-1 tumors. The diffusion time was 560.5 ms.

Animal Number	Intercept	Slope	r-value
1	18 (2)	0.94 (0.01)	0.95
2	6 (2)	0.93 (0.01)	0.95
3	4 (1)	0.90 (0.01)	0.99
4	-35 (3)	1.15 (0.02)	0.93
5	23 (6)	0.93 (0.04)	0.77
6	-28 (8)	1.17 (0.08)	0.65
7	2 (2)	0.98 (0.02)	0.95
8	28 (4)	0.47 (0.04)	0.69
9	-6 (1)	1.07 (0.01)	0.98
10	40 (6)	0.72 (0.06)	0.53
11	-50 (11)	1.5 (0.1)	0.69
Mean (SEM)	0 (2)	0.98 (0.01)	

An example of the diffusion data used in this study is represented in Fig. 4.3. The solid line schematically represents the behavior of $D(t)$ as $t^{1/2}$ is varied. $D(t)$ is plotted versus $t^{1/2}$ since in that representation, the slope of the curve is proportional to S/V for short diffusion times. Four maps, representative of different regimes along the $D(t)$ curve, were chosen for further study from the 20 calculated maps. The diffusion times of these four maps were 11.0, 58.0, 360.5, and 560.5 ms. These maps are representative of: 1) the short time regime (or S/V regime) in which $D(t)$ is proportional

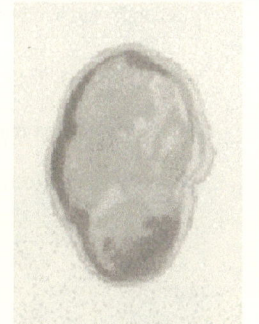

Fig. 4.2. Example of a histological slide used to determine necrotic regions for RIF-1 tumors. This slide is for the tumor whose pO_2 maps are shown in Fig. 4.1. Light area indicate viable tissue.

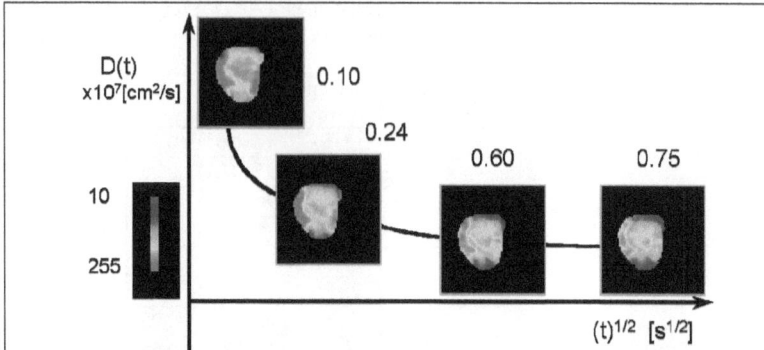

Fig. 4.3. Schematic representation of a typical $D(t)$ versus $t^{1/2}$ curve for a RIF-1 tumor showing the four maps used in the analysis (out of the twenty acquired). Numerical labels on the diffusion maps are the $t^{1/2}$ in s^{-1} (diffusion times range from 11.0 to 560.5 ms). The color scale represents diffusion coefficients from 0.10×10^{-5} cm^2/s to 2.55×10^{-5} cm^2/s. Note that the regions display different dependencies on the diffusion time, i.e., the $D(t)$ for the central region (associated with necrotic tumor tissue) changes little with diffusion time, while the periphery (associated with viable tumor tissue) generally has a larger time dependence.

to S/V (11.0 ms), 2) the transition regime in which $D(t)$ switches from the S/V regime to the effective media regime (58.0 ms), 3) the "near" effective media regime (360.5 ms), and 4) the "far" effective media regime in which the diffusion time is long enough such that all of the tissue is in the effective media regime and $D(t)$ is proportional to $1/\tau$ (560.5 ms).

Fig. 4.4 shows an example of the scatter plots for the shortest and longest diffusion times studied of $D(t)$ for the animal breathing air versus $D(t)$ measured during carbogen breathing. The solid lines are least-squares fit to the data. The mean slopes and intercepts for all 11 animals are presented in Table 4.2. In all animals the data scatter decreased as the diffusion time was increased.

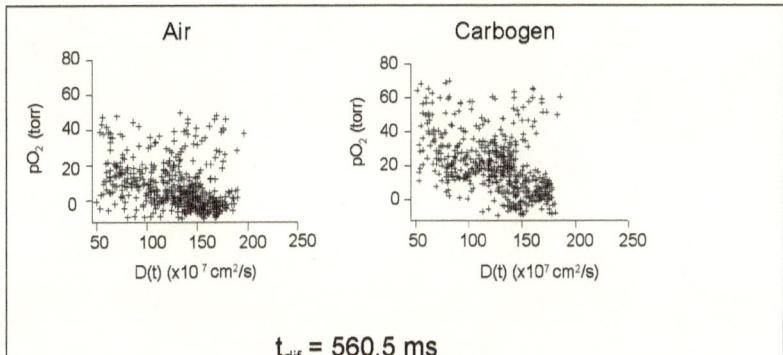

t_{dif} = 560.5 ms

Fig. 4.4. Scatter plots for the shortest and longest diffusion times showing the correlation between $D(t)$ for air breathing and $D(t)$ for carbogen breathing for a single RIF-1 tumor. $D(t)$ for short diffusion times reflects local properties of the environment such as the ratio of surface are to volume, S/V. At longer times, $D(t)$ is indicative of "effective" properties of the medium, such as the tortuosity, τ. The solid lines in each plot are linear least-squares fits to the data. The fit parameters are: for t_{diff} = 11.0 ms, slope = 0.92 ±0.02, intercept = -1 ± 4 (r = 0.86) and for t_{diff} = 560.5 ms, slope = 0.93 ± 0.01, intercept = 6 ± 2 (r = 0.95).

Scatter plots of $D(t)$ versus pO_2 for a single RIF-1 tumor are shown in Fig. 4.5 for both air and carbogen breathing. While there is significant scatter in the data, the bulk of the data in both plots with a high value of $D(t)$ has a relatively low pO_2. The $D(t)$ map with the longest diffusion time (560.5 ms) was used to generate these example plots, but the same behavior was seen for all diffusion times used in this study. The effect of a change from air to carbogen breathing gas is obvious in Fig. 4.6: pixels with values at the lower end of the range of $D(t)$ values are the most affected in pO_2 value. This was the case for each animal in this study. No attempt was made to determine mean fit parameters from all the data since the slope of each line is determined by the change in pO_2 value for each pixel and the degree of change varied significantly for each tumor. However, to give the reader some indication of the statistics associated with these data, the fit parameters for

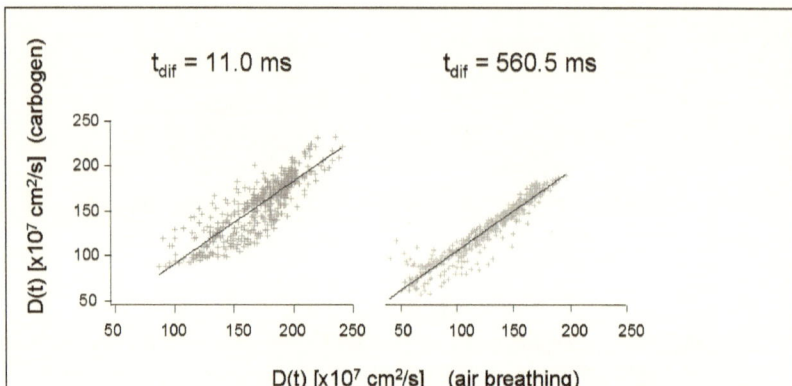

Fig. 4.5. Scatter plots for air and carbogen breathing showing the correlation between $D(t)$ and pO_2 for the longest diffusion time for a single RIF-1 tumor. These data confirm that carbogen breathing affects only the well vascularized tumor periphery (associated with the lower $D(t)$ values) and to a smaller extent the tissue in the necrotic center of the tumor (associated with the higher $D(t)$ values.

the data shown in Fig. 4.5 were: slope = -0.01 ± 0.01, intercept = 23 ± 1, Pearson's r = 0.54 (air breathing) and slope = -0.01 ± 0.01, intercept = 38 ± 2, Pearson's r = 0.48 (carbogen breathing).

One of the difficulties with the above presentation is that it is not clear how an individual pixel's $D(t)$ and pO_2 values change following a change in breathing gas. In order to examine this issue more carefully, two quantities were calculated: $\Delta D(t)$ (=$D(t)\_c - D(t)\_a$, where $D(t)\_c$ and $D(t)\_a$ are the diffusion coefficients measured with the animal breathing carbogen and air, respectively) and ΔpO_2 (=$pO_2\_c - pO_2\_a$, where the pO_2_c and pO_2_a are the oxygen tension values measured with the animal breathing carbogen and air, respectively). Shown in Fig. 4.6a is a plot of ΔpO_2 versus $\Delta D(t)$ for one representative tumor. This plot allows the identification, on a pixel-by-pixel basis, of how a change in pO_2 is reflected as a change in $D(t)$. The shape of the distribution of pixel values can be obtained from projection of the data onto each axis. These projections are shown as histograms in Figs. 4.7b and 4.7c. The distribution for $\Delta D(t)$ is roughly normal and centered around zero, i.e., on average there is no net change in $D(t)$ for a change in breathing gas. The histogram for pO_2, as expected, does reflect an increase with a change to carbogen breathing.

4.1.6. Discussion

Diffusion-coefficient mapping has been shown to be a useful too in distinguishing pathology from normal tissue in many applications (Moseley *et al.*, 1990; Knight *et al.*, 1991; Helmer *et al.*, 1995). By exploiting the structural changes that often accompany pathology, *ADC* mapping can aid both in its visualization and in the determination of tissue types. ^{19}F NMR of sequestered PFC emulsions has been shown to be a rapid and quantitative method of mapping tumor pO_2 distributions in murine tumors (Baldwin and Ng, 1992; Dardzinski and Sotak, 1994). Together, these two methods allow for a unique view of tumor tissue oxygenation and a method for testing any possible relationship between *ADC* and pO_2.

In agreement with a previous study using perfluoro-15-crown-5-ether (Dardzinski and Sotak, 1994) and studies using other PFCs (Parhami and Fung, 1983; Kong *et al.*, 1984;

Clark et al., 1984; Reid et al., 1985; Sotak et al., 1993), the R_1 relaxation rate was found to be linearly related to dissolved oxygen concentration and temperature. This relationship was used to calculate *in vivo* pO_2 maps on a pixel-by pixel basis. Shown in Fig. 4.1 are examples of these maps. These maps have the same slice thickness as the *ADC* maps and thus are an improvement over the projection images previously obtained using this PFC (Dardzinski and Sotak, 1994). The pO_2 values for the air breathing animal (Fig. 4.1a and 4.1b) are the largest in the periphery of the tumor where the vasculature is presumably intact. The lowest pO_2 values were found in the center of the tumor, a region that displayed evidence of necrosis (as determined from histological data). A histological slice for the tumor in Fig. 4.1 is shown in Fig. 4.2. The colors have been reversed for the greatest contrast, and hence light areas correspond to viable tissue. Note that the viable tissue areas correspond well to the most well-oxygenated regions in Fig. 4.1.

The distribution of pO_2 values is determined by the final location of the PFC within the tumor. The PFC is delivered to the tumor through fenestrations in the vasculature (Ratner *et al*, 1988), the distribution of pO_2 values found in these experiments will be weighted toward higher pO_2 values. This is due to the fact that, to reach less well-perfused or necrotic regions, the PFC will either have to diffuse to those regions, or an initially well-perfused region may become hypoxic as the tumor grows during the time allowed for the PFC to clear the vasculature. None of the pO_2 maps in this study exhibited regions in which there was no signal from the PFC. This is most likely a result of the volume averaging inherent in the 3 mm slice thickness used in these measurements.

The range of pO_2 values in the histograms in Fig. 4.1 is taken from –20 to 80 torr. The precision of the R_1 measurement resulted in a precision in pO_2 of ± 5 torr, consistent with earlier studies (Dardzinski and Sotak, 1994). The occurrence of negative pO_2 values is most likely due to the assumption that the tumor temperature is 37°C. In this experiment the animals' body temperature is maintained by a flow of 37°C air initially, reduced to 34°C after 10 minutes. This reduction is necessary to prevent hyperthermia in the animal. Because the tumor is located on the back of the animal and has a large surface area, it is likely that the tumor temperature is not equal to the core temperature, and is somewhere between 34°C and 37°C. In addition, the compromised circulation between body and tumor impedes a major source of heat equilibration in the body. According to the calibration equation, a reduction in temperature results in a reduction in pO_2, by approximately 3 torr/°C. The negative pO_2 values are consistent with the precision of these experiments (± 5 torr) and a tumor temperature decreased from the core temperature. This offset in pO_2 is, however, of little consequence in the present experiments because any correlation between ADC and pO_2 would be independent of the offset. In addition, since the measurement performed with carbogen breathing are compared with those in the same animal breathing air, any offset in pO_2 will be cancelled when differences are taken. The range of positive pO_2 values found in this experiment are consistent with needle electrode measurements performed by Terris et al. (1992), which found values up to 60 torr in RIF-1 tumors with air breathing.

Hypoxic cells are thought to play an important role in the resistance of solid tumors to radio- and chemotherapy. Carbogen breathing is known to increase the radiosensitivity of these hypoxic cells in murine tumors (Suit et al., 1972; Siemann et al., 1977) and to increase the pO_2 only in particular locations in the tumor (Dardzinski and Sotak, 1994). In this study carbogen breathing was used to change the tissue oxygenation in order to explore any concomitant changes in $D(t)$. By changing the oxygen tension distribution in the tumor, and additional test can be made as to the correlation between $D(t)$ and pO_2. For example, if $D(t)$ and pO_2 appears to be correlated in a particular region, but $D(t)$ does non increase as pO_2 increases, this correlation can be determined to be false or coincidental. Data for carbogen breathing presented in Fig. 4.1c shows that the largest increases in pO_2 are confined to the periphery. It is in this region that any correlation between $D(t)$ and pO_2 would be expected as it includes viable, as well as hypoxic, tissue. The necrotic regions are not expected to have much variation in the value of $D(t)$ or pO_2, and therefore any correlation might be weaker in these regions.

A related issue is the determination of the diffusion time that yields optimal differentiation between viable, hypoxic, and necrotic tissue. By optimizing the diffusion time, the dynamic range of $D(t)$ can be maximized and a clearer evaluation of the correlation between $D(t)$ and pO_2 will result. The *ADC* of water molecules diffusing in

RIF-1 tumors has been shown to be time dependent (Helmer et al., 1995). In the present experiment, this is shown schematically in Fig. 4.3. In Fig. 4.3, large values dominate the $D(t)$ map at short diffusion times, both in the central regions and in the periphery. As the diffusion time is increased, the majority of pixels in the periphery show a decline in $D(t)$. In the central region, however, the diffusion coefficient shows only a small decline with increasing diffusion time. This indicates how $D(t)$ maps can be used to differentiate between areas of different tumor tissue types: $D(t)$ maps can be acquired for a range of diffusion times and the time-dependent behavior can indicate the different regions. The difference between time-dependent behaviors is maximized at long diffusion times and it is these diffusion times, therefore, that will most aid in the differentiation between necrotic and viable tumor tissue.

In order to illustrate the effects of diffusion time on the correlation plots of $D(t)$ for air breathing versus $D(t)$ for carbogen breathing, the extreme cases (diffusion times of 11.0 and 560.5 ms) are presented in Fig. 4.4. It is clear from the two plots that the diffusion time influences the spread in the data. The fit parameters are: for t_{diff} = 11.0 ms, slope = 0.93 ± 0.02, intercept = -1 ± 4 (r = 0.86) and for t_{diff} = 560.5, slope = 0.93 ± 0.01, intercept = 6 ± 2 (r = 0.95). The increase in the correlation coefficient with diffusion time is consistent with the idea that the measured diffusion coefficient, at longer diffusion times, reflects the longer scale structure of the sample and not simply the local variations. It may be argued that the shortest diffusion time within the effective-media regime would be the best choice for analysis since that would minimize the averaging over different tissue types. The longest diffusion time of 560.5 ms was used for analysis, however, since it corresponds to a diffusion length of 33 μm, or roughly an order of magnitude smaller than the pixel length of 470 μm (the voxel size is 470 μm x 470 μm x 3 mm) and therefore, the partial volume averaging arising from the pixel size will dominate the effects due to the diffusion time. In addition, the fact that the slope is close to unity and intercepts are close to zero implies that carbogen breathing has little, if any, effect on $D(t)$.

Table 4.2 presents the slope and intercept values for each of the measured tumors as well as the average for each parameter. The small average intercept (0 ± 2, mean ± SEM) as well as the average slope near unity (0.98 ± 0.01, mean ± SEM) implies that the change in $D(t)$ brought about by carbogen breathing is a small effect at most. This data has also been analyzed using the Restricted Maximum Likelihood (REML) method (Laird and Ware, 1982) which iteratively estimates the random variances of the slopes and intercepts of these data for each animal. The variances are used as weights for the original data points (pixel values) in a weighted least-squares fit. The REML analysis gave, for the entire data set, slope = 0.97 ± 0.03 and intercept = 1±2. The p-value for the intercept being different from zero was 0.72 and the p-value for the slope being different from unity was 0.30 and, therefore, neither value was statistically different. This result is interesting in that, while it may be expected that the correlation between $D(t)$ and pO_2 is rather weak in the viable tissue that is already well-oxygenated, there is no population of pixels that exhibits a large shift in $D(t)$. This restricts the usefulness of using $D(t)$ as a clinical indicator of pO_2, since the sensitivity to relatively small pO_2 changes seems to be low. While a number of animals individually displayed statistically significant deviations from the null result, these deviations did not correlate with other parameters measured in this study such as tumor volume. Therefore, the results for all animals taken together are reported.

Although there is no statistical correlation between $D(t)$ and pO_2, as the data presented in Fig. 4.5 shows, there are a number of interesting relationships between the data in different regions of these plots. The scatter plots emphasize that the greatest change in pO_2 is for pixels with the lowest values of $D(t)$, pixels that were identified with reasonably well-vascularized, non-necrotic tissue in Figs. 1 and 3. The pixels with the highest values of $D(t)$ (identified as necrotic tissue by H & E staining) show little if any change in pO_2 with change in breathing gas. There has been no attempt to make a linear fit to these data for each animal as the slope with air breathing is highly dependent upon tumor size and, therefore, necrotic fraction. The change in slope with breathing gas is also highly variable for the same reason, since only the pO_2 of viable tissue is affected by a switch to carbogen breathing. To illustrate the degree of correlation that is obtained from these plots, the data in Fig. 4.5 were subjected to a linear least-squares fit (pO_2 = intercept + (slope)(ADC)) with these results shown as a solid line in both cases. For the animal breathing air, the intercept = 23 ± 1 and slope = -0.012 ± 0.009 (Pearson's r = 0.54), while for carbogen breathing, the intercept = 38 ± 2 and slope = -0.014 ± 0.012 (r = 0.48).

The data suggests the following picture: hypoxic regions in tumors become necrotic as the tumor volume increases and cells become further removed from the tumor vasculature. As necrosis proceeds, cell membranes rupture and the resulting debris is subsequently degraded by auto- or heterolysis. It would be expected that water diffusing in viable tissue and higher *ADCs* are therefore expected for necrotic regions. Consequently, for lower *ADCs* (corresponding to viable tumor tissue) greater values of pO_2 are expected. Lower (or zero) pO_2 values are associated with necrotic tissue that have correspondingly higher *ADC* values.

In contrast to the above results, Dunn *et al.* (1995) found a *positive* correlation between *ADC* and pO_2 in a study combining NMR *ADC* maps with oxygen tension measured using EPR of implanted LiPc crystals. Oxygen tension measurements were performed in the region of the pixels with the highest and lowest values on an *ADC* map. The results in that study were assumed to hold only for non-necrotic regions. In the present work, all pixels (with sequestered PFC) are included in the analysis, and presumably include viable, hypoxic, and necrotic tissue. With reference to Fig. 4.5, it can be seen that even when the pixels at the highest *ADC* values ($> \approx 1.4 \times 10^{-5}$ cm^2/s and corresponding to necrotic tissue) are excluded, the correlation between $D(t)$ and pO_2 is still non-existent. In addition, given the broad spread in the pixel data, there are many possible choices of two *ADC* values at the extreme ends of the range that would demonstrate a positive correlation between *ADC* and pO_2. Unfortunately, this approach does not capture the complexity of the data and can lead to erroneous conclusions.

In summary, this study demonstrates the absence of correlation between $D(t)$ and pO_2 in RIF-1 tumors when viable, hypoxic, and necrotic tissue are all included. Furthermore, excluding necrotic tissue data still results in no correlation between tumor water *ADC* and pO_2 as measured using PFCs. However, this method, which combines both $D(t)$ and pO_2 measurements, may be useful in following treatment regimens and for establishing treatment efficacy in a noninvasive manner. Changes in tumor tissue viability and oxygen status can be imaged through the use of sequestered PFCs and necrotic tissue can be separated from viable tissue using $D(t)$ maps. This study also demonstrates that the best $D(t)$ contrast between necrotic and non-necrotic tissue is achieved at long diffusion times ($>\approx 100$ ms in the RIF-1 model).

4.1.7. Acknowledgements

The authors thank R. J. Kaufman, Ph.D. and HemaGen/PFC (St. Louis, MO) for providing the perfluoro-15-crown-5-ether used in this study. The authors also acknowledge Gail Boulienne of the University of Massachusetts Medical Center for her excellent histological work. The authors also thank Jeff Dunn for useful discussions relating to Fig. 4.6. Joseph D. Petruccelli from the department of Mathematical Sciences of Worcester Polytechnic Institute (WPI) is thanked for assistance in regards to the REMP analysis. David S. Adams of the Biology and Biotechnology Department of WPI is thanked for the use of his imaging system to digitize the histological slides. Part of this work is supported by a Biomedical Engineering Research Grant from The Whitaker

Foundation (K. G. H.). Part of this work was performed during the tenure of an Established Investigatorship from the American Heart Association (C. H. S.).

Chapter 5

Yeast Experiments

The goal of these experiments was to characterize the diffusion behavior of water in a two-compartment system. Yeast cell suspensions have been well characterized by others and are fairly uncomplicated model systems to understand. The following work contains multiple sets of experiments using yeast cells as a model system to characterize the diffusion behavior of water.

5.1. Deconvolution of Restriction Effects on Compartmental Diffusion Using Combined Relaxometric and Diffusimetric NMR

5.1.1. Abstract

Diffusion signal attenuation curves in yeast-cell suspensions show non-monoexponential signal decay which is assumed to arise from separate compartmental contributions to the overall signal. However, restricted diffusion effects also give rise to non-monoexponential signal decay and are difficult to separate from compartmental signal contributions. Combined relaxometry and diffusion measurements allows differentiation between compartmental diffusion constants by first separating the compartmental contributions on the basis of differences in their respective relaxation times. Diffusion-weighted inversion-recovery spin-echo experiments were carried out at different b-values. Intra- and extracellular compartments in yeast-cell suspensions were separated on the basis of T_1 relaxation by adding an MR contrast agent to the extracellular space. Once the compartmental signals were distinguished on the basis of T_1, the relative signal attenuation for each compartment was used to calculate the separate compartmental ADCs. With this method, even compartmental diffusion coefficients with similar values can be distinguished.

5.1.2. Introduction

The cell membrane plays an important role in cell physiology by acting as a protective barrier between the extra- and intracellular compartments. Since the cellular membrane is a lipid bilayer structure with a highly hydrophobic core (Fox, 1993), it is unclear how water is transported between the intra- and extracellular spaces through a seemingly impermeable membrane. It has been speculated that transmembrane transport of water occurs *via* osmo-regulated water channels (Jap and Li, 1995; Skach *et al.*, 1994) or in conjunction with water-soluble ion transport.

Movement of water between the extra- and intracellular spaces, as well as within the cells, is of particular importance in biomedicine. Studies have shown that changes in the apparent diffusion coefficient (ADC) of water in biological systems are a function of membrane permeability and changes in the extracellular volume fraction (Latour et al., 1993b, 1994). Since the water NMR signal is measured simultaneously from the intra- and extracellular space, it is difficult to ascertain whether the changes in the ADC are caused by drastic changes in one or both of the compartments, or by permeability changes in the cell membrane.

Non-monoexponential behavior in the diffusion-signal attenuation curves has been observed in a number of biological systems and has been attributed to compartmental effects (Niendorf et al., 1996; Andrasko, 1976b; Vétek et al., 1994; Szafer et al., 1995). Although it has been reported that changes in the extracellular volume fraction mirror the changes observed in the rapidly decaying component of the signal decay curve (Trouard et al., 1997; Niendorf et al., 1994), the volume fraction calculated for the slowly diffusing component in these experiments differ significantly from the true intracellular volume fraction (Niendorf et al., 1994). While multi-compartment systems in the slow-exchange regime are expected to exhibit non-monoexponential signal decay curves, Helmer et al. (1995) have shown that non-monoexponential diffusion-signal attenuation curves can be obtained in polystyrene bead packs (a single compartment system) due to restriction effects alone. Consequently, since the compartmental contributions do not definitively represent the cellular compartments, it is difficult to ascertain whether the non-monoexponential behavior in a highly restricted two-compartment system is due to compartmental contributions, restrictions, or both of these effects.

Since compartmental effects and membrane permeability are closely related, comparable methods have been used to study both parameters (Andrasko, 1976a; Tanner, 1983; Pilatus *et al.*, 1997). Several studies have shown the efficacy of using MR contrast agents to discriminate between extra- and intracellular water signals (Labadie *et al.*, 1994; Stanisz *et al.*, 1998), or to measure membrane permeability (Andrasko, 1987a; Conlon and Outhred, 1972). MR contrast agents do not cross the cell membrane and thus selectively affect the extracellular water signal. The addition of MR contrast agents to the extracellular space drastically alters the relaxation properties of the sample, and, with sufficient concentrations of contrast agent, the system can be brought into the slow exchange regime where a biexponential relaxation curve characterizes the separate contributions from the intra- and extracellular compartments (Labadie *et al.*, 1994).

Another method used to discriminate between intra- and extracellular water signals is by adding diffusion gradients to a Carr-Purcell Meiboom-Gill (CPMG) sequence (van Dusschoten *et al.*, 1996; Herbst and Goldstein, 1989). A major disadvantage of using the transverse relaxation to distinguish between compartments is that T_2 is sensitive to bulk magnetic susceptibility effects, diffusion, and exchange. Kleinberg *et al.* (1994) have shown that T_2 relaxation times in porous media are reduced by diffusion in the presence of inhomogeneous magnetic fields arising from magnetic susceptibility differences, while T_1 is dominated by relaxivity due to hyperfine interactions with the pore surface. While these studies were performed on porous rocks, the mechanism of relaxation is very similar for water in the presence of paramagnetic MR contrast agents.

In previous experiments (Han *et al.*, 1998a, 1998b), T_1 relaxation was used to separate the compartmental signal contributions using relaxography. This method effectively differentiates the compartments on the basis of T_1 relaxation, which then allows other types of measurements to be performed on the separate compartmental signals (Labadie *et al.*, 1994; Provencher and Dovi, 1979). These works also show that if the continuous relaxation spectra display two distinct distributions of relaxation times, comparable results can be obtained by fitting the data to a biexponential using a constrained nonlinear least-squares algorithm (Han *et al.*, 1998a; Silva *et al.*, 1998).

In this study, combined T_1 relaxation and diffusion data were obtained from yeast-cell suspension in the presence of a MR contrast agent. From these data, the effective compartmental ADCs were calculated and compared to the decay constants calculated from a biexponential fit of the diffusion-signal attenuation curve. The validity and reliability of such a model to determine compartmental ADCs is discussed.

5.1.3. Theory

It has been shown that the relationship between a set of relaxation-signal decay data and the probability distribution of relaxation times is given by the Laplace transform. Thus, by taking the formal inverse Laplace transform (ILT) of data governed by a distribution of relaxation times

$$M_{xy} = \sum_i \left[M_{0i} \left(1 - 2e^{-\frac{T_I}{T_{1i}}} \right) \right]$$ [5.1.3.1]

where M_{xy} is the NMR signal and TI is the inversion time, the relaxogram generates a spectrum of relaxation time constants, T_{1i}, and their respective contributions to the overall signal, M_{0i} (Labadie *et al.*, 1994). It has also been shown that even for relaxograms displaying only a single distribution, the corresponding relaxographic images from different sections of the distribution can represent spatially distinct populations of spins (Pályka *et al.*, 1994; Dardzinski *et al.*, 1994).

The two-compartment exchange equation has been analytically solved by Kärger *et al.* (1988), where the normalized signal for an NMR inversion-recovery experiment is given by the equation

$$M_{norm}(t) = \frac{M_{xy}(t)}{M_0} = M'_{0a}\left(1 - 2e^{-\frac{t}{T_{1a}}}\right) + M'_{0b}\left(1 - 2e^{-\frac{t}{T_{1b}}}\right)$$ [5.1.3.2]

where

$$M'_{0b} = \frac{1}{2}\left\{1 - \frac{\left[(M_{0b} - M_{0a})\left(\frac{1}{T_{1a}} - \frac{1}{T_{1b}}\right) + \frac{1}{\tau_a} + \frac{1}{\tau_b}\right]}{\sqrt{\left(\left(\frac{1}{T_{1b}} - \frac{1}{T_{1a}}\right) + \left(\frac{1}{\tau_b} - \frac{1}{\tau_a}\right)\right)^2 + \frac{4}{\tau_a \tau_b}}}\right\} \qquad [5.1.3.3]$$

$$M'_{0a} + M'_{0b} = 1 \qquad [5.1.3.4]$$

$$\frac{1}{T'_{1a}} = \frac{1}{2}\left[\left(\frac{1}{T_{1b}} + \frac{1}{T_{1a}} + \frac{1}{\tau_b} + \frac{1}{\tau_a}\right) - \sqrt{\left(\left(\frac{1}{T_{1b}} - \frac{1}{T_{1a}}\right) + \left(\frac{1}{\tau_b} - \frac{1}{\tau_a}\right)\right)^2 + \frac{4}{\tau_a \tau_b}}\right]$$

$$[5.1.3.5]$$

$$\frac{1}{T'_{1b}} = \frac{1}{2}\left[\left(\frac{1}{T_{1b}} + \frac{1}{T_{1a}} + \frac{1}{\tau_b} + \frac{1}{\tau_a}\right) + \sqrt{\left(\left(\frac{1}{T_{1b}} - \frac{1}{T_{1a}}\right) + \left(\frac{1}{\tau_b} - \frac{1}{\tau_a}\right)\right)^2 + \frac{4}{\tau_a \tau_b}}\right]$$

$$[5.1.3.6]$$

where M'_{0a} and T'_{1a} are the fast decaying component, and M'_{0b} and T'_{1b} are the slow decaying component of the signal; T_{1a}, T_{1b}, M_{0a}, and M_{0b} are the actual compartmental relaxation times and volume fractions; τ_a and τ_b are the mean lifetimes (or residence times) of the spins in their respective environments.

It is evident that when the system is in slow exchange (i.e., $\tau_a, \tau_b \to \infty$), the fast and slow decaying components of relaxation approximate the fast and slow relaxing compartments ($T'_{1a} \to T_{1a}$ and $T'_{1b} \to T_{1b}$), and the signal-attenuation curve exhibits a biexponential decay. Conversely, when the system is in fast exchange ($\tau_a, \tau_b \to 0$), the dominating effect is the exchange rate ($\frac{1}{\tau_a}$ and $\frac{1}{\tau_b}$) and the signal-attenuation curve exhibits a monoexponential decay.

For a two-compartment system displaying a single relaxographic distribution, the system is in fast exchange and the single distribution represents the weighted average of both the intra- and extracellular water signals. Adding MR contrast agent to the extracellular space effectively acts as a shift reagent in relaxographic space. As the concentration of MR contrast agent increases, the width of the distribution will broaden and shift to lower values of relaxation times. With increasing concentrations of contrast agent, the extracellular T_1 will become the dominating term in Eqs. [5.1.3.5] and [5.1.3.6] and the system will shift from the fast- (residence-time dominated) to the slow-exchange (relaxation-time dominated) regime. At some concentration of extracellular MR contrast agent, there will be a complete separation of the relaxographic distributions, where the fast component of relaxation will be centered around T'_{1a} and the slow component around T'_{1b}. It is important to note that the T'_{1a} and T'_{1b} are not the actual compartmental relaxation times, but rather, the effective compartmental relaxation times (the dominating effects on T'_{1a} and T'_{1b} are T_{1a} and T_{1b}, respectively). Similarly, M'_{0a} and M'_{0b} are the effective compartmental volume fractions.

It is a nontrivial matter to extract M_{0a} and M_{0b} from M'_{0a} and M'_{0b} since *a priori* knowledge of the residence times as well as one of the relaxation time constants is necessary. A reliable estimate of the residence times can be obtained from (Conlon and Outhred, 1972; Andrasko, 1976b; Kärger *et al.*, 1988)

$$\tau_b = \frac{1}{P}\left(\frac{V}{A}\right) \qquad [5.1.3.7]$$

$$\tau_a = \tau_b \left(\frac{M_{0b}}{M_{0a}}\right) \qquad [5.1.3.8]$$

where P is the membrane permeability, and $\frac{V}{A}$ is the ratio of the volume to surface area of the cells. Since the residence times are a function of the compartmental ratios, at least one of these quantities must be measured independently (Herbst and Goldstein, 1989).

It can be shown that for a system in slow exchange, a constrained nonlinear least-squares fit to Eq. [5.1.3.2] is comparable to results obtained using CONTIN (Han *et al.*, 1998b).

COMPARTMENTAL DIFFUSION

Taking this one step further, Eq. [5.1.3.2] can be modified (Eqs [5.1.3.9] and [5.1.3.10]) to include echo time (*TE*) effects, gradient (*b*) effects, and the inversion efficiency (M_{offset}).

$$M_{norm}(T_I, TE, b) = \frac{M_{xy}(T_I, TE, b)}{M_0} \qquad [5.1.3.9]$$

$$M_{norm}(T_I, TE, b) = M'_{0a}\left(1 - 2e^{-\frac{T_I}{T_{1a}'}}\right)e^{-\frac{TE}{T_{2a}'}}e^{-bD_a'} + M'_{0b}\left(1 - 2e^{-\frac{T_I}{T_{1b}'}}\right)e^{-\frac{TE}{T_{2b}'}}e^{-bD_b'} + M_{offset}$$

$$[5.1.3.10]$$

where the variables represented denote the effective time constants and fractional contributions (Fig. 5.1.4.1). It should be noted that while the T_1 and T_2 terms are sensitive to the presence of MR contrast agent, the diffusion coefficient values are relatively unaffected by such paramagnetic substances. Hence, the compartmental differentiation on the basis of relaxometry in the presence of MR contrast agents will have little effect on the subsequent measurements of the compartmental diffusion coefficients.

Although exchange effects are present, the spin densities associated with each of the relaxographic peaks represent the medium effectively sampled by those spins during the echo time. Hence, for systems in slow exchange, $M_{0a}^{'}$ represents the components of relaxation dominated by spins in the fast-decaying compartment (M_{0a}), while $M_{0b}^{'}$ represents the components of relaxation dominated by spins in the slow-decaying compartment (M_{0b}). Furthermore, despite the pulse sequence used, spins will effectively experience the same medium as long as the evolution times (time interval between excitation and acquisition) are identical.

5.1.4. Methods

5.1.4.1. Yeast Preparation

Baker's dry yeast (Fleishmann's Yeast, Inc., Oakland, CA), 1.5 g, was rehydrated at room temperature in a 50 cc centrifuge tube by adding 40 cc of distilled H_2O. The suspension was bubbled with medical-grade air and, after a starving period of approximately 3 hrs, the suspension was centrifuged for 8 min at 3500 rpm (IEC-Centra8 Centrifuge, International Equipment Company, USA). Yeast cells were washed four times with 5 mM gadopentate dimeglumine (Gd-DTPA, Magnevist®, Berlex, Wayne, NJ). After the final wash, the wet/dry mass ratio was adjusted to 3.25:1 and NMR experiments were performed. The elapsed time between the end of the sample preparation and the beginning of NMR data acquisition was about ten minutes.

5.1.4.2. NMR Experiments

Combined Relaxometry and Diffusion (CRD)

All data were obtained on a GE-CSI-II 2.0T/45 cm imaging spectrometer with ±20 G/cm self-shielded gradients operating at 85.5 MHz for ^{1}H. The studies were performed using a 5-mm-diameter, 9-turn solenoid coil. The magnet-bore temperature was 19°C. NMR data were acquired using an inversion-recovery pulse sequence with diffusion-weighted spin-echo detection (Fig. 5.1.4.1). Thirty-two inversion times (TI), with logarithmic temporal spacing ($TI = TI_0 * a^{n-1}$, $TI_0 = 2.0$ ms, $a = 1.250$) and two signal averages, were used to measure each T_1-relaxation curve. Measurements were repeated for eighteen different b-values (1410, 5639, 12688, 22557, 35245, 50754, 69082, 90229, 114196, 140983, 170590, 203016, 238262, 276327, 317212, 360917, 407442, and 456786 s/cm^2). Other acquisition parameters were $\Delta = 41.5$ ms, $\delta = 6.0$ ms, pre-delay = 2000 ms. Half-sine-shaped gradient pulses with amplitudes of 1.7, 3.5, 5.2, 6.9, 8.7, 10.4, 12.1, 13.9, 15.6, 17.3, 19.1, 20.8, 22.5, 24.2, 26.0, 27.7, 29.4, and 31.2 G/cm were used for the diffusion weighting. Since the maximum gradient strength was ±20 G/cm for each axis, gradients were applied simultaneously in three orthogonal directions to achieve the necessary gradient values. Half-echoes were acquired with a spectral width of ±1250 Hz and 2K data points. Separate measurements were performed at $TE = 55$ and 65 ms to characterize echo-time effects.

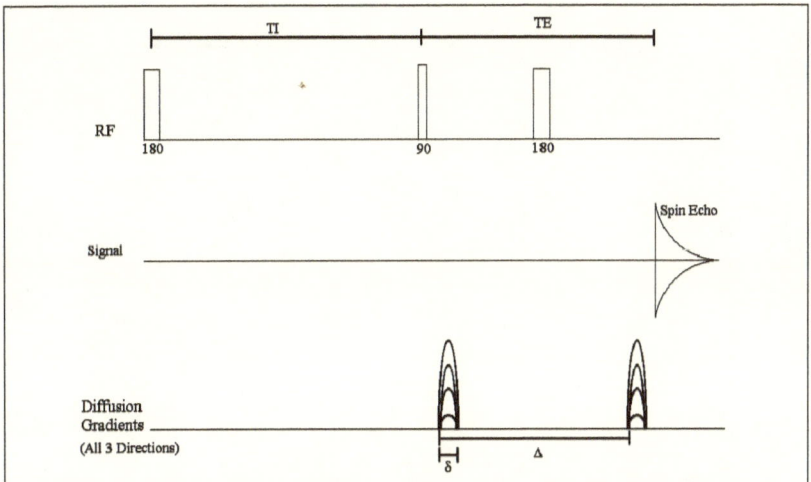

Fig. 5.1.4.1. Inversion-recovery spin echo pulse sequence with diffusion gradients. TI ranged from 2 to 2019 ms in logarithmic increments where $TI = TI_0 * a^{n-1}$, $TI_0 = 2.0$ ms, $a = 1.250$, and $n = 0...31$, $TE = 55$ ms, $g = 1.73$ to 31.2 G/cm in 1.73 G/cm increments, $\Delta = 41.5$ ms, $\delta = 6.0$ ms, and b-values of 1410, 5639, 12688, 22557, 35245, 50754, 69082, 90229, 114196, 140983, 170590, 203016, 238262, 276327, 317212, 360917, 407442, and 456786 s/cm². Gradients were applied in all three orthogonal directions simultaneously to achieve these b-values.

Pulsed-Field-Gradient Spin-Echo (PFGSE) Experiment

To examine the behavior of the signal as a function of b-value, a standard pulsed-field-gradient spin-echo (PFGSE) sequence (Fig. 5.1.4.2; Stejskal and Tanner, 1965) with 32 b-values, $TE = 55$ ms, $\Delta = 41.5$ ms, $\delta = 6.0$ ms, and two signal averages was used. The b-values were varied by keeping the diffusion time constant and increasing the gradient strength from 0.97 G/cm to 31.04 G/cm, in increments of 0.97 G/cm. Gradients were applied simultaneously in three orthogonal directions in order to achieve these gradient values.

COMPARTMENTAL DIFFUSION

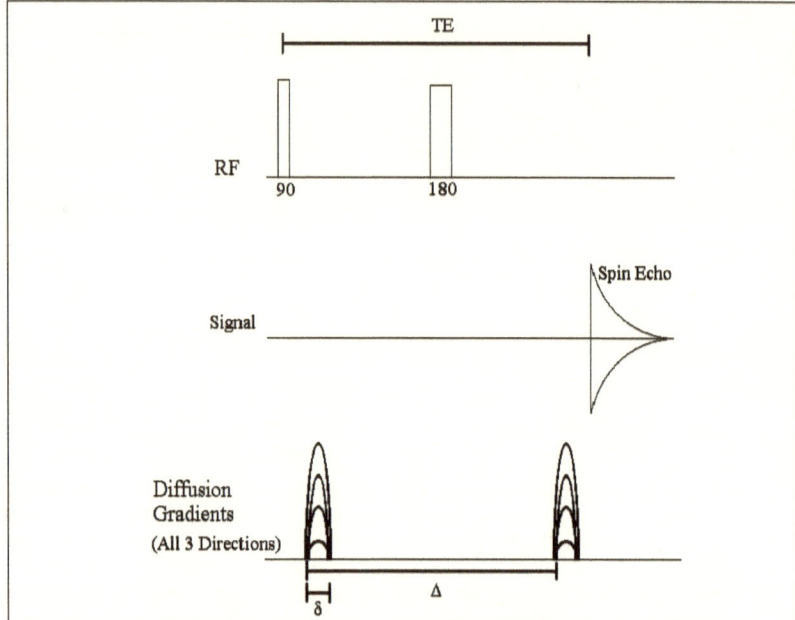

Fig. 5.1.4.2. Pulsed-field-gradient spin-echo (PFGSE) pulse sequence. The diffusion time was kept constant and the *b*-value varied by incrementing the total gradient strength from 0.97 G/cm to 31.04 G/cm. Gradients were applied simultaneously in all three orthogonal directions. Other acquisition parameters were $TR = 2000$ ms, $TE = 55$ ms, $\Delta = 41.5$ ms, and $\delta = 6.0$ ms.

5.1.4.3. Data Analysis

Inversion-Recovery Data

Data were evaluated using CONTIN (Provencher, 1982a; 1982b; Provencher and Dovi, 1979) and also by constrained-biexponential fitting to the equation

$$M_{xy} = M_{0a}\left(1-2e^{-\frac{T_I}{T_{1a}}}\right) + M_{0b}\left(1-2e^{-\frac{T_I}{T_{1b}}}\right) + M_{Offset} \qquad [5.1.4.1]$$

where M_{xy} represents the measured signal intensity, M_{0a} and T_{1a} are the M_0 and T_1 of the fast-decaying component of the signal, M_{0b} and T_{1b} are M_0 and T_1 of the slow-decaying component of the signal, and finally $M_{0Offset}$ represents the inversion efficiency. A standard nonlinear least-squares algorithm available in IDL (Research Systems, Inc., Boulder, Colorado) was used for the constrained-biexponential fitting. The T_1 and M_0 values obtained from the constrained-biexponential fit were compared for consistency to results obtained using CONTIN. The resulting M_{0a} and M_{0b} at each b-value were then used to calculate the respective ADCs of both the fast- and slow-relaxing component of the overall signal.

In a similar fashion, the echo-time dependence of the signal was studied for the two sets of inversion-recovery curves generated at $TE = 55$ and 65 ms. The inversion-recovery data were fitted to Eq. [5.1.4.1] and the resulting M_{0a} and M_{0b} at each echo time were fitted to a two-point T_2 measurement according to Eq. [5.1.4.2].

$$T_2(ms) = \frac{(65-55)}{\ln\left(\frac{M_{xy_{55}}}{M_{xy_{65}}}\right)} \qquad [5.1.4.2]$$

This yielded an approximation for the T_2 relaxation times of the fast and slow decaying signals.

PFGSE Data

The diffusion attenuation curves generated by the PFGSE sequence were fitted to a biexponential diffusion-attenuation curve using

$$M_{xy} = M_{0a}e^{-bD_a} + M_{0b}e^{-bD_b} \qquad [5.1.4.3]$$

where M_{0a} and M_{0b} are the calculated fast and slow components of M_0 for the biexponential decay, and D_a and D_b are their respective decay constants.

5.1.5. Results

5.1.5.1. [Gd-DTPA] and Relaxographic Peak Separation

Relaxography data for yeast-cell suspensions in 3.0 mM Gd-DTPA shows that there are clearly two distinct components, and is consistent with the results obtained by Labadie *et al.* (1994). At 2.0T and 19°C, the T_1 of 3.0-mM Gd-DTPA in the bulk solution was measured to be 60 ms.

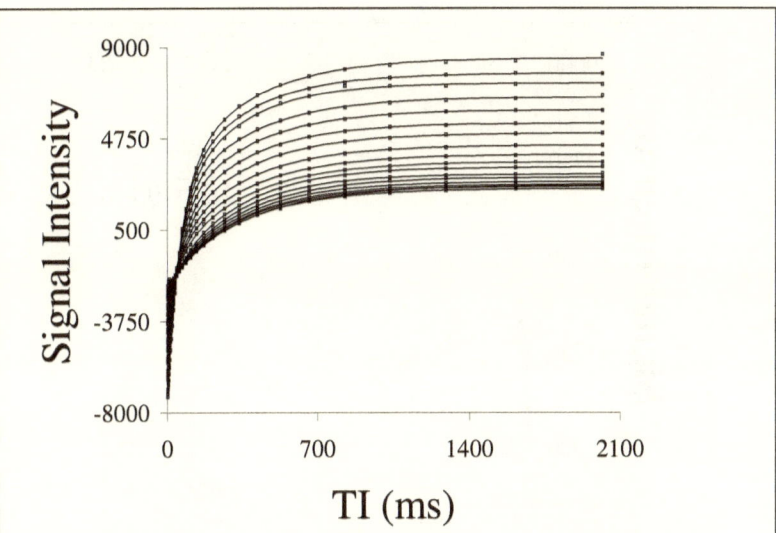

Fig. 5.1.5.1. Inversion-recovery curves at different gradient values. Inversion-recovery data acquired with the following NMR parameters: TI ranged from 2 to 2019 ms in logarithmic increments where $TI = TI_0 * a^{n-1}$, $TI_0 = 2.0$ ms, $a = 1.250$, and $n = 0...31$, $TE = 55$ ms, $g = 1.73$ to 31.2 G/cm in 1.73 G/cm increments, $\Delta = 41.5$ ms, $\delta = 6.0$ ms, and b-values of 1410, 5639, 12688, 22557, 35245, 50754, 69082, 90229, 114196, 140983, 170590, 203016, 238262, 276327, 317212, 360917, 407442, and 456786 s/cm². All of these curves display biexponential behavior and were fitted to

$$M_{xy} = M_{0a}\left(1-2e^{-\frac{TI}{T_{1a}}}\right) + M_{0b}\left(1-2e^{-\frac{TI}{T_{1b}}}\right) + M_{Offset}$$

The solid lines on the plot represent the fit to the data while the points represent the actual data.

5.1.5.2. Inversion-Recovery Experiments with Varying Gradient Strengths and Echo Times

The b-value-dependent inversion-recovery curves and the fit to the curves (Fig. 5.1.5.1) show an excellent fit of the data to Eq. [5.1.4.1]. The calculated T_1 and M_0 values for the

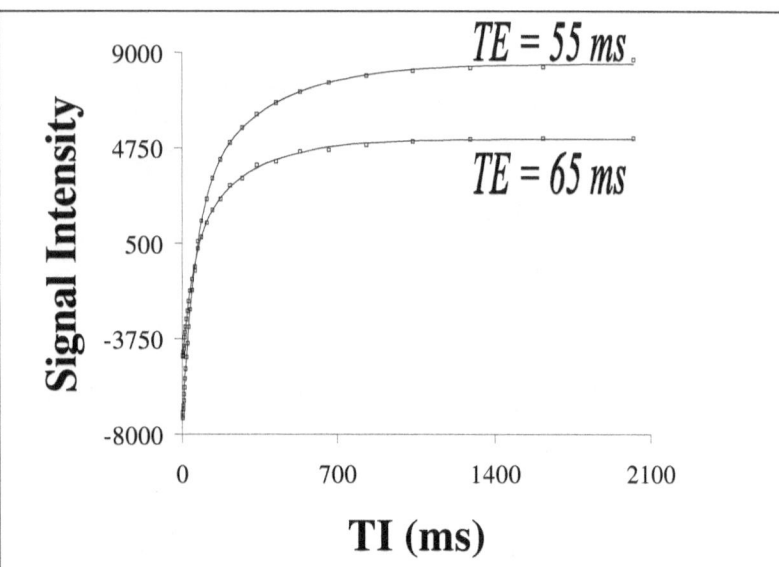

Fig. 5.1.5.2. Inversion-recovery curves at different echo times. Inversion-recovery data were acquired with the following NMR parameters: $TI = TI_0 * a^{n-1}$, $TI_0 = 2.0$ ms, $a = 1.128$, $TE = 55$ and 65 ms. Both curves display biexponential behavior and were fitted to

$$M_{xy} = M_{0a}\left(1-2e^{-\frac{T_I}{T_{1a}}}\right) + M_{0b}\left(1-2e^{-\frac{T_I}{T_{1b}}}\right) + M_{Offset}$$

The solid lines on the plot represent the fit to the data while the points represent the actual data

fast and slow components of decay at different b-values are summarized in Table 5.1.5.2.1. The b-value-dependent behavior of M_{0a} exhibits a larger attenuation than M_{0b}. This indicates that M_{0a} is associated with the faster-diffusing extracellular spins while M_{0b} is associated with slower-diffusing intracellular spins.

Table 5.1.5.2.1. Calculated T_1 and M_0 values for the fast and slow components of relaxation at different b-values. T_{1a} and M_{0a} represent the effective relaxation time and spin density for the fast decaying extracellular compartment while T_{1b} and M_{0b} represent the effective relaxation time and spin density for the slow decaying intracellular compartment at a 55 ms echo time. No T_1 value was available for the fit of the inversion recovery curve to the highest b-value. This indicates a quenching of the extracellular compartment at this b-value.

b-value (s/cm²)	M_{0a}	T_{1a} (ms)	M_{0b}	T_{1b} (ms)
1410	4855	59	3182	328
5639	4310	54	3171	296
12688	3780	54	3188	284
22557	3587	57	2697	333
35245	2986	55	2720	320
50754	2519	55	2663	310
69082	2127	57	2580	324
90229	1668	56	2531	319
114196	1363	57	2430	331
140983	1007	56	2462	325
170590	561	43	2709	305
203016	405	24	2642	294
238262	317	52	2478	318
276327	205	37	2424	313
317212	185	55	2309	326
360917	118	52	2302	321
407442	233	83	2173	368
456786	134	***	2275	319

The TE-dependent inversion-recovery curves also exhibit biexponential behavior. The calculated M_0 and T_1 values associated with the fast and slow components of relaxation are summarized in Table 5.1.5.2.2. The estimates of T_2 relaxation times for M_{0a} and M_{0b} were 16 ms and 34 ms for the fast and slow components, respectively.

Table 5.1.5.2.2. Calculated T_1 and M_0 values for the fast and slow components of relaxation at different echo times. T_{1a} and M_{0a} represent the effective relaxation time and spin density for the fast decaying extracellular compartment while T_{1b} and M_{0b} represent the effective relaxation time and spin density for the slow decaying intracellular compartment at 55 ms and 65 ms echo times.

	TE55	TE65
M_{0a}	4855	2600
T_{1a}	59 ms	53 ms
M_{0b}	3182	2367
T_{1b}	328 ms	257 ms

5.1.5.3. Calculation of the Biexponential Diffusion-Decay Constants from the PFGSE Data

The non-monoexponential diffusion-signal attenuation curve obtained using the PFGSE sequence is shown in Fig. 5.1.5.3. The curve exhibits biexponential behavior and was fitted to Eq. [5.1.4.3] to yield

$$M_{xy} = 0.436 e^{-1.64 \times 10^{-5} b} + 0.564 e^{-1.75 \times 10^{-6} b} \qquad [5.1.5.1]$$

There was excellent correspondence between the data and the fit as can be seen in Fig. 5.1.5.3.

5.1.5.4. Calculation of ADCs Associated with the Fast and Slow Relaxing Components of the Signal

The dependence of M_{0a} and M_{0b} on b-value is shown in Fig. 5.1.5.4. The behavior of both M_{0a} and M_{0b} exhibit non-monoexponential attenuation. When fitted to a single exponential, M_{0a} (the component associated with the extracellular space) showed a

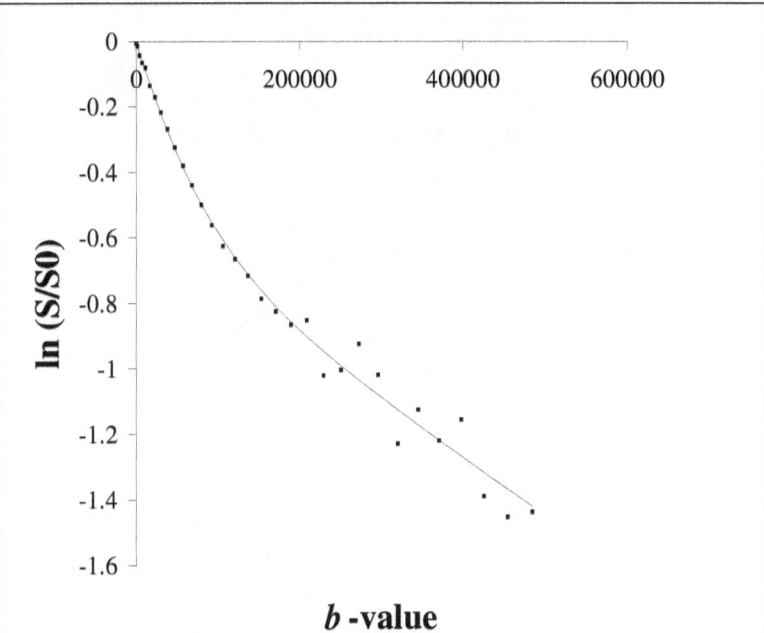

Fig. 5.1.5.3. Diffusion attenuation curve for yeast-cell suspension acquired using the PFGSE pulse sequence in Fig. 5.1.4.2. Non-monoexponential behavior in diffusion attenuation can be seen in yeast samples. The diffusion attenuation curves were acquired with the following NMR parameters: $TE = 55$ ms, $\Delta = 41.5$ ms, $\delta = 6.0$ ms, $g = 0.48*n$ G/cm where $n = 1\ldots32$. The data were fitted to a biexponential decay curve

$$M_{xy} = 0.436e^{-1.64\times10^{-5}b} + 0.564e^{-1.75\times10^{-6}b}$$

The solid line on the plot represents the fit to the data while the points themselves are the actual data

reasonably good fit while M_{0b} (component associated with the intracellular space) produced a poorer fit. The fits to a monoexponential attenuation function were

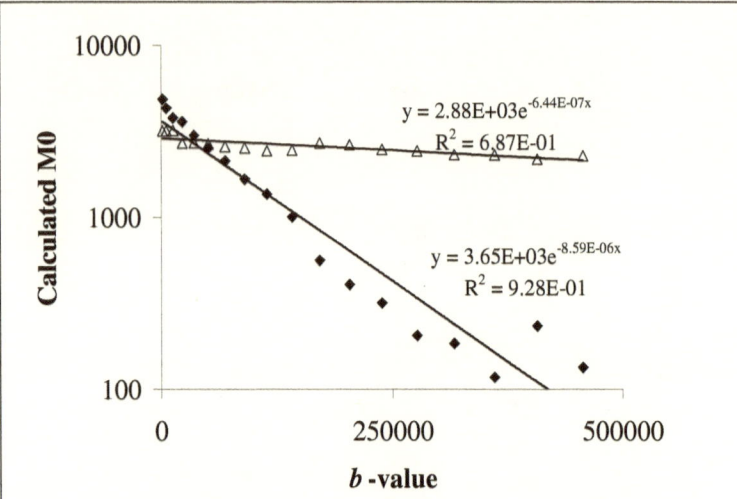

Fig. 5.1.5.4. Calculated M_{0a} and M_{0b} from water in yeast-cell suspensions plotted as a function of b-value. Both curves display non-monoexponential diffusion attenuation. The M_0 values were fitted to

$$M_{xy_{fast}} = 0.559e^{-8.59\times10^{-6} b}; \qquad R^2 = 0.928$$

for the fast component and

$$M_{xy_{slow}} = 0.441e^{-6.44\times10^{-7} b}; \qquad R^2 = 0.687$$

for the slow component.

$$M_{xy_{fast}} = 0.559e^{-8.59\times10^{-6} b}; \qquad R^2 = 0.928 \qquad [5.1.5.2]$$

and

$$M_{xy_{slow}} = 0.441e^{-6.44\times10^{-7} b}; \qquad R^2 = 0.687 \qquad [5.1.5.3]$$

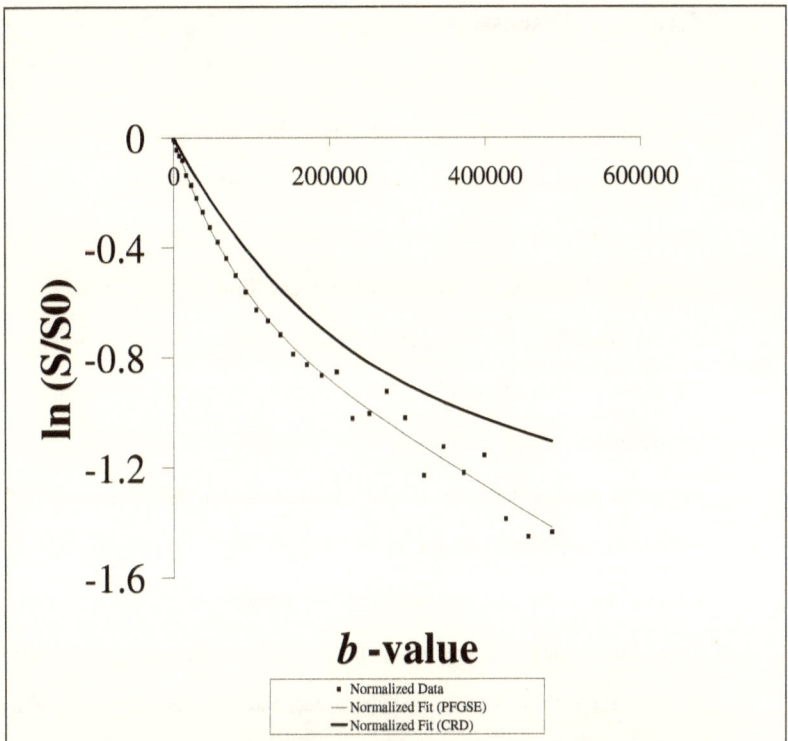

Fig. 5.1.5.5. Normalized plot of the calculated fits using the two different methods. PFGSE fit to the biexponential is displayed along with the two fits obtained from the combined relaxography and diffusion (CRD) fits. Both the monoexponential and biexponential fit to the calculated values of M_0 are shown. It is evident that all three curves exhibit distinct behavior.

There was a significant discrepancy between this method and the PFGSE sequence when the ADC value was calculated for the fast-decaying compartment.

5.1.6. Discussion

Differentiation of compartmental signals using MR contrast agents has been shown to be effective for yeast-cell suspensions. For this particular system, where a complete separation of relaxographic peaks was observed, it is believed that the measured M_0 values are dominated by compartmental rather than exchange effects. It was also observed that fits of M_{0fast} versus b-value to a monoexponential produced good results while the corresponding fit to M_{0slow} produced a much poorer result. Since the compartmental effects are thought to be deconvolved using relaxometry, there is no theoretical rationale for fitting the calculated M_0 values to a biexponential. The larger errors associated with the fits for the intracellular ADC could partly be due to the reduced dynamic range for the attenuation of the intracellular signal as compared to the extracellular signal. Another possibility is that the intracellular water experiences an environment with greater restrictions, hence, the diffusion attenuation exhibits non-monoexponential behavior. In either case, it becomes necessary to be able to measure M_{0fast}, M_{0slow}, T_{1fast}, and T_{1slow} to a high degree of accuracy. Furthermore, if indeed the intracellular water displays a non-monoexponential decay due to restriction, more b-values would be necessary to observe this effect.

The calculated T_2 values from the echo-time dependent inversion-recovery curves show that there is significant signal loss when a 55 ms echo time is used. In both the extra- and intracellular compartments, there is a greater than 80% signal attenuation due to the echo time used for this experiment. More importantly, since the measured T_2 values were not dissimilar enough, it would have been difficult to differentiate the compartments on the basis of T_2 alone. One of the advantages of using T_1 is that the underlying mechanism for spin-lattice relaxation is the hyperfine interaction of the ^1H_2O protons with the paramagnetic center of the MR contrast agent (Kleinberg et al., 1994). This avoids complications arising from through-space bulk-susceptibility effects that can affect the T_2 relaxation time of water in compartments that have no MR contrast agent present.

The PFGSE data exhibit non-monoexponential diffusion-signal attenuation behavior. Although an excellent fit to this data can be obtained using a biexponential, the interpretation of these data becomes problematic since part of the curvature most likely arises from restriction effects. Since the intracellular compartment is highly restricted ($\approx$4-μm-diameter yeast cells; Tanner, 1983), the restriction effects are probably non-negligible at the diffusion time and gradient strengths used for this experiment. For this reason, the discrepancy between the PFGSE biexponential fit and the CRD fit for the data was not surprising. Due to the relatively low packing density of the yeast-cell suspension, restriction effects in the extracellular space are probably minimal. This hypothesis is substantiated by the monoexponential signal decay of the deconvolved extracellular component. However, there is a significant discrepancy between the fits for the intracellular ADC between the PFGSE and CRD methods. We believe this disparity is due to restriction effects. Since the M_{xy} and ADC values measured from the PFGSE experiment are a representation of the environment sampled by the water molecules during the evolution time, it is possible that the space sampled during this echo time is a combined function of exchange and restrictions. While the sampled space would be the same for the CRD experiment, the compartmental contributions are deconvolved from the inversion-recovery data before the ADC is calculated. This allows for independent calculations of ADC from each compartment without the worry of contamination due to exchange effects.

One of the effects observed in this study was the eventual quenching of the extracellular signal at high b-values. This can account for the missing value of T_1 measured for the inversion-recovery curve at the highest b-value. This idea of quenching the signal could prove to be useful in studies where only a single compartment is of interest.

In order to understand the restriction effects on the overall signal, it would be useful to study the time-dependent behavior of water in a restricted two-compartment system. Investigation of the time-dependent behavior of compartmental water would provide valuable information on the microstructure of the environment that the ^1H_2O samples (Helmer *et al.*, 1995; Latour *et al.*, 1993a; Norris and Niendorf, 1995).

A major advantage of using combined relaxometry with diffusion (CRD) gradients is that two different compartments with similar ADC values can be distinguished on the basis of differences in relaxation times of the respective compartments.

In conclusion, we have shown that compartmental contributions in the measurement of ADC can be deconvolved by combined relaxometry and diffusimetry. This method unequivocally differentiates between the intra- and extracellular signals before fitting for the ADC, thus removing the confounding effects of restriction on the measured ADC values and also allowing for ADC measurements of two compartments with similar ADC values.

Chapter 6

Rabbit Achilles Tendon Experiments

This chapter is a pre-print of the article summarizing initial experiments by Stephen Gemmell and the continuation of that work. One of the problems encountered after the first set of experiments was that the tendon water *ADC* was drastically affected by the storage medium. Due to osmotic imbalances, the phosphate-buffered saline caused the tendon to swell. For this reason, it was uncertain whether the changes in *ADC* were due to tendon characteristics, or rather, if these changes were an artifact of the storage medium. This finding led to the use of tendons from freshly sacrificed rabbits and NMR measurements were obtained immediately post sacrifice. While the trend in the *ADC* with load was similar to that of the swollen tendon, other characteristics were quite different. These experiments were conducted with Dr. Peter Grigg. My role in all of these experiments was to acquire and process the data, and also to write the NMR methods and analysis of the results. Stephen Gemmell and Dr. Peter Grigg were responsible for the preparation and mounting of the swollen tendon and the freshly harvested tendon, respectively.

6.1. Characterization of Water *ADC* Behavior Under Tensile Loading and Recovery in Rabbit Achilles Tendon Using NMR

S. S. Han[1], S. J. Gemmell[1], K. G. Helmer[1], A. H. Hoffman[2], P. Grigg[3], and C. H. Sotak[1,4]

[1] Department of Biomedical Engineering, Worcester Polytechnic Institute, Worcester, MA 01609
[2] Department of Mechanical Engineering, Worcester Polytechnic Institute, Worcester, MA 01609
[3] Department of Physiology, University of Massachusetts Medical School, Worcester, MA 01605
[4] Department of Radiology, University of Massachusetts Medical School, Worcester, MA 01605

Running Title: NMR Measurements of Water Transport in Tendons

Address correspondence to:
    C. H. Sotak
    Department of Biomedical Engineering
    Worcester Polytechnic Institute
    100 Institute Road
    Worcester, MA 01609
    Tel: 508 831 5617
    Fax: 508 831 5541
    email: csotak@wpi.edu

6.1.1. Abstract

Water diffusion measurements were made on rabbit Achilles tendon to determine their behavior during static tensile loading and unloading. Tendons previously stored frozen in phosphate-buffered saline (PBS) were sequentially loaded with 0.4-, 5-, 10-, and 0.4-N loads. The apparent diffusion coefficient (ADC) was measured perpendicular ($ADC_\perp$) and parallel ($ADC_\parallel$) to the fiber orientation at diffusion times of 10, 30, and 60 ms for each load. $ADC_\perp$ and $ADC_\parallel$ increased with increasing load for all samples. New samples of freshly-harvested tendons were loaded with a static 5-N load for five minutes and then unloaded for 30 minutes to determine the effects of loading and unloading. The $ADC_\perp$ of fresh tendon was studied as a function of loading and unloading. The $ADC_\perp$ increased with load for all samples. This increase was attributed to the extrusion of tendon water into a bulk phase outside the tendon. The recovery of the tendon upon unloading exhibited a reversal of the $ADC_\perp$ back to the baseline value. This recovery was attributed to the water moving from the bulk phase to the bound phase. The recovery followed a slower time course than the extrusion of water. It was also found that the phosphate-buffered saline caused the tendon to swell. This method can be used both to detect structural changes in tendon under tensile loading and to study the transport of water in tendon.

6.1.2. Introduction

The structure and behavior of connective tissue is of great importance to biomechanics and rehabilitation engineering. There is a significant amount of evidence to show that the properties of these connective tissues are strongly influenced by their hydration characteristics. It has been shown that the soft-tissue characteristics change with water content. Furthermore, the water content changes with mechanical loading of the soft tissue (Burstein *et al.*, 1993). When cartilage is subjected to compression, there is bulk transport of internal water; the viscous drag associated with this flow through the solid matrix of the tissue is a major determinant of its viscoelastic behavior (Mow *et al.*, 1984). Movement of water has also been shown to be caused by tensile loading of soft tissue (Hannafin and Arnoczky, 1994, Chimich *et al.*, 1992). These investigators showed that tensile loading led to a reduction in water content of tendons and ligaments. It has been shown that water removal causes stiffening (Betsch *et al.*, 1980; Eldon, 1964; Galante, 1967), while increasing hydration results in softening (Viiidik *et al.*, 1966; Eldon, 1964) of tissues loaded in tension. Furthermore, the viscous response of ligaments is altered when the water content is changed (Chimich *et al.*, 1992). Thus, water movement may play a significant role in determining the material properties of soft tissue structures under tensile load.

The characterization of water behavior in soft tissues under tensile loads has been limited due to the measurement methods previously employed. Hannafin & Arnoczky (1994) and Chimich *et al.* (1992) subjected tissue to tensile loads and determined water content using destructive measures. Those methods made it impossible to make within-sample comparisons or real-time observations. We have used nuclear magnetic resonance (NMR) methods to characterize water in tensile loading of rabbit Achilles tendon. The measured apparent diffusion coefficient (*ADC*) of water can reflect the local or global tissue structure experienced by the diffusing water molecules and can be used to detect structural changes (Helmer *et al.*, 1995). Water confined to the interior of the tendon should have a lower *ADC* than water in the bulk phase due to the restrictions posed by the collagen.

6.1.3. Methods

6.1.3.1. Equipment and Apparatus

All tendons were mounted vertically (perpendicular to the long axis of the magnet bore) in a section of capillary tubing with a 5.3-mm inner diameter. The tendon and tube were placed on a mounting apparatus consisting of a simple pulley system that allowed loads to be applied exterior to the magnet. A 9-turn solenoid radiofrequency (RF) coil that conformed to the outside of the glass tubing was used for excitation and detection.

All experiments were performed on a GE CSI-II 2.0T/45 cm imaging spectrometer operating at 85.56 MHz for protons and equipped with ±20 G/cm self-shielded gradients.

All tendons were immersed in paraffin oil to prevent dehydration (Beaulieu and Allen, 1994). The pulse sequences used were modified for paraffin-oil NMR-signal suppression to include chemical-shift-selective binomial pulses (Hore, 1983). The frequency separation between the water and paraffin resonances was 312.5 Hz.

6.1.3.2. Load Dependence of Tendon Water *ADC*

Studies were conducted on rabbit Achilles tendon (n=10), of approximately 25-mm length, and having cross-sectional areas of ~7 mm^2. Tendons were harvested from rabbits sacrificed in unrelated research. Individual tendons were separated and stored in phosphate-buffered saline (pH = 7.4) at −30°C. The tendons were thawed at the time of the experiment, mounted in the apparatus, and NMR measurements were initiated immediately.

Upon securing the sample to the apparatus, a 10-N load was applied for ≈5 min to ensure the integrity of the attachment. This procedure also constituted the preconditioning protocol and established a pseudoelastic state in the tendon. Once secured and preconditioned, a tendon was allowed to recover in an unloaded state for approximately 10 min. It was then subjected to experimental loading. Sequential loads of 0.4, 5, 10, and 0.4 N were applied to each tendon. The 0.4-N load was applied to remove slack from the tendon and establish a reference state. After the application of each load, a period of 15 min was allowed to elapse before NMR data were acquired. Preliminary experiments showed that the ADC of tendon water reached an equilibrium state after a 5-min loading period. Thus, the tendons were assumed to be at equilibrium after 15 min of load. For each load, $ADC_\perp$ and $ADC_\parallel$ were measured at 10, 30, and 60 ms diffusion times (t_{diff}). A pulsed-field-gradient spin-echo (PFGSE, Stejskal and Tanner, 1965) sequence with the following parameters was used to measure $ADC_\perp$ and $ADC_\parallel$: two signal averages, TR = 1000 ms, δ = 6.0 ms, and TE = 120 ms to ensure that the same population of spins was being observed. These experiments conformed to a 2 (directions) × 3 (diffusion times) × 4 (loads) experimental design utilizing repeated measures, and the results were analyzed using a 2 × 3 × 4 analysis of variance (ANOVA) with repeated measures.

6.1.3.3.   Diffusion-Time Dependence of the ADC

$ADC_\perp$ and $ADC_\parallel$ were measured in the tendons previously stored in phosphate-buffered saline. A 0.4-N load was applied to the tendons to remove slack and ADC measurements were taken using a PFGSE sequence with similar NMR acquisition parameters as the load dependent experiment. The ADC was measured at 14 diffusion times, ranging from 8.0 to 100.0 ms. Both the standard PFGSE and a spin-echo bipolar gradient-pulse (BGP, Trudeau *et al.*, 1995; Hong and Dixon, 1992) diffusion-weighted sequence were used to cover the range of diffusion times. The BGP sequence employed a bipolar pair of diffusion-weighted gradient pulses on either side of the 180° RF pulse and was used for diffusion times between 8.0 and 40.0 ms. The PFGSE sequence was used to collect data with diffusion times of 60.0 – 100.0 ms.

6.1.3.4. Fast-Time Measurements of Water *ADC* in Response to Uniaxial Tensile Loading

Fresh tendons were used for this set of experiments to avoid artifacts suspected to be caused by storage in phosphate-buffered saline. Tendons were harvested immediately post mortem in a chamber with 100% humidity, immersed in paraffin oil, and NMR data were acquired. Upon securing the sample to the apparatus, a 0.4-N load was applied to remove slack from the tendon. The $ADC_\perp$ was measured using six gradient strengths (3, 6, 9, 12, 15, and 18 G/cm), four signal averages, $\delta = 6.0$ ms, TE = 25.0 ms, pre-delay of 1700 ms, and $t_{diff} = 80.0$ ms. Data were collected at one-minute intervals. Five minutes of baseline data were collected and then a 5-N static load was applied to the tendon and data were collected for five minutes. The 5-N load was removed and data were collected for 35 minutes at the 0.4-N load (Fig. 6.1). Immediately following collection of the last $ADC_\perp$ data set, an $ADC_\parallel$ data set was collected in order to deterimine if there was directional anisotropy of the *ADC* value.

The baseline value for the $ADC_\perp$ measurements was calculated as the average of the five pre-load data sets. Phenomenological models based on experimental results have shown that the mechanical behavior of the tendon closely resembles that of a parallel combination of Maxwell (damped spring) and Hooke (spring). This combined effect represents the response of a Voight solid to a sudden load, and has been previously used (Fung, 1993; Frisen *et al.*, 1969) to model the behavior of elasto-mechanical solids and follows the equation described by

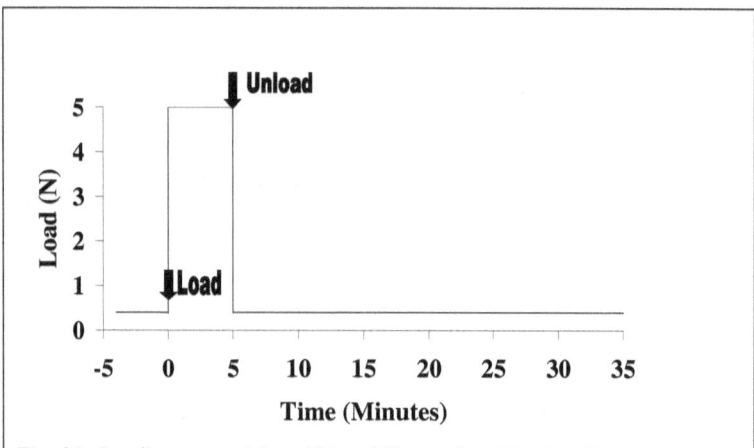

Fig. 6.1. Loading protocol for rabbit Achilles tendon. Five baseline sets collected at 0.4 N. 5-N load at t=0 – 5 minutes. Unloaded to 0.4-N at 5-minute time point and data collected for additional 30 minutes.

$$c(t) = \frac{1}{\mu}\left(1 - e^{-\frac{\mu}{\eta}t}\right)u(t) \qquad [6.1]$$

where $c(t)$ is the response to a unit step function, $u(t)$, with viscosity, η, and spring constant, μ. Since the behavior of the tendon ADC on loading was speculated to follow a similar function, the ADC response was fitted to such a step response curve following the equation

$$\%Change_{rise} = \%_{max}\left(1 - e^{-\frac{t}{T_{rise}}}\right) \qquad [6.2]$$

for the time points $t = 0 - 5$ minutes, where $\%_{max}$ is the extrapolated equilibrium value to which the $ADC_\perp$ would rise, and T_{rise} is the rise-time constant. Since the tendon ADC behavior is presumed to be correlated to the mechano-elastic behavior of the tendon, such a model should adequately represent the behavior of ADC in relation to load.

For the fall-time calculations, a standard decaying exponential of the form:

$$\%Change_{fall} = \%_{start} e^{-\frac{t-5}{T_{fall}}} + \%_{offset} \qquad [6.3]$$

was used, where T_{fall} represents the fall time. In some cases, the tendon $ADC_\perp$ never recovered to the baseline value, and thus the equation was adjusted to compensate for such an offset by adding a constant value, $\%_{offset}$, to the fitting function. The rise and fall time constants were calculated from the average percentage changes of all tendons (n = 8) in this set of experiments.

6.1.3.5. Studies of Tendon Mechanical Behavior

In separate experiments, not utilizing NMR, tendons were studied in a tensile loading apparatus in which both stresses and strains could be determined under conditions similar to the NMR experiments. These experiments were performed to determine the stress-strain curve for tendons under similar loading. Tendons were secured at both ends using suture material. The sutures were secured to clamps coupled to a linear actuator and a load cell in an apparatus that has been described in detail elsewhere (Duquette et al., 1996). A microscope was mounted above the tendon so that the behavior of the tendon could be observed during loading and its diameter could be measured optically. Stresses were calculated from the applied loads and the cross-sectional area. Strains were determined by measuring the displacements of small markers fixed on the tendon surfaces.

6.1.4. Results

6.1.4.1. Load Dependence of Tendon Water *ADC*

The effect of loading on tendons stored in PBS was studied in experiments where $ADC_\parallel$ and $ADC_\perp$ were measured at three diffusion times for each of the four loading conditions. Both $ADC_\perp$ and $ADC_\parallel$ increased as the load was increased from 0.4 N to 10 N (Fig. 6.2). The effect of loading on the measured *ADC* was highly significant ($P<0.0001$). Recovery was not complete at the time of measurement of the second 0.4-N load. In addition, the $ADC_\perp$ was significantly less than the $ADC_\parallel$ ($ADC_\parallel / ADC_\perp = 1.3$ at t_{diff} of 8 ms, $ADC_\parallel / ADC_\perp = 1.7$ at t_{diff} of 100 ms; $P<0.0001$). Also, the load-dependent change in $ADC_\perp$ was greater than the load-dependent change in $ADC_\parallel$.

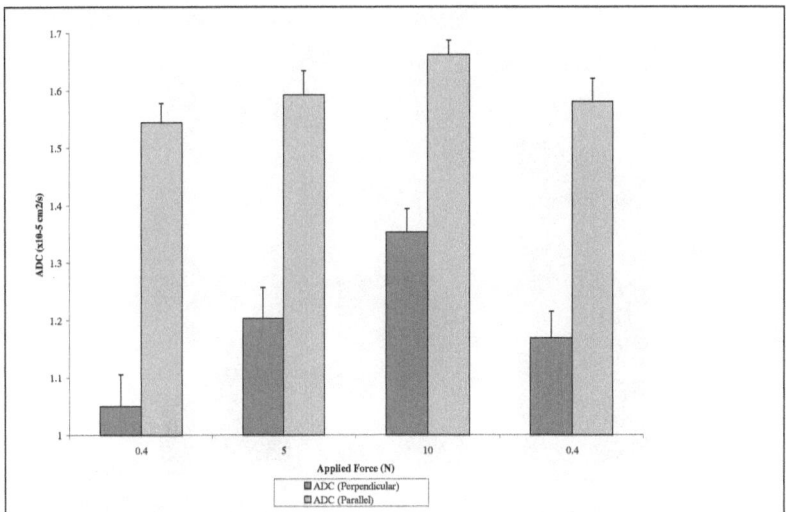

Fig. 6.2. Load dependence of $ADC_\parallel$ and $ADC_\perp$ for applied loads of 0.4, 5, 10, and 0.4 N. Error bars are ± SEM. The increase in ADC with load for both directions is attributed to the extrusion of bulk water that forms an envelope around the tendon. The smaller increase in $ADC_\parallel$ with increasing load is thought to be due to the fact that the water molecules are already relatively free to diffuse in that direction and the addition of the signal from the bulk extruded water therefore has only a small effect.

6.1.4.2. Diffusion-Time Dependence of the *ADC*

In order to thoroughly explore the effect of diffusion time, the *ADC* of tendons stored in PBS was measured in separate trials with a nominal 0.4-N load. Diffusion times were varied from 8.0 to 100.0 ms (Fig. 6.4). Diffusivity decreased along both directions as a function of the time allowed for diffusion. The effect of diffusion time was highly significant in these experiments ($P<0.0001$).

TENDON

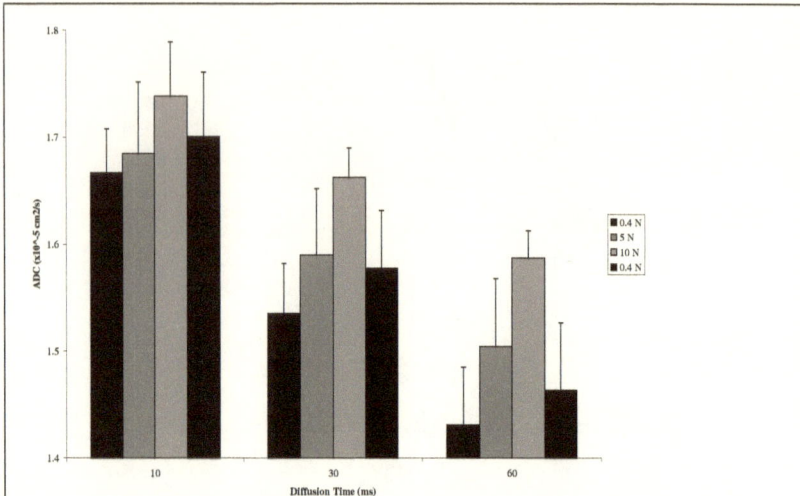

Fig. 6.3. Load dependence of $ADC_\perp$ for applied loads of 0.4, 5, 10, and 0.4 N at 10, 30, and 60 ms diffusion times. Error bars are ± SEM. The increase in ADC with load is attributed to the extrusion of bulk water that forms an envelope around the tendon. Larger changes associated with longer diffusion times are attributed to restriction effects.

Information about characteristic structural sizes can be obtained from Fig. 6.4. At short diffusion times it has been shown (Mitra *et al.*, 1992; Helmer *et al.*, 1995) that the behavior of the *ADC* vs. $t^{1/2}$ is proportional to the ratio of the surface to volume (*S/V*) of the medium in which the water molecules are diffusing (Eq. [6.4]):

$$\frac{ADC}{D_0} = 1 - \frac{4}{9\sqrt{\pi}} \frac{S}{V} \sqrt{D_0 t} \qquad [6.4]$$

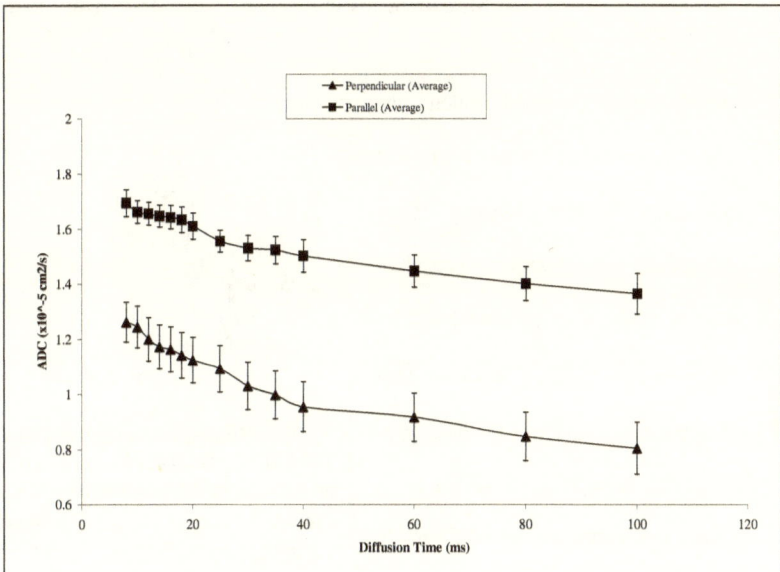

Fig. 6.4. The diffusion-time dependence of the water $ADC_\perp$ and $ADC_\parallel$. Error bars are ± SEM (n=10). Note that the initial slope of the curves are different for the two directions, signaling a difference in the ratio of surface area to volume.

where D_0 is the diffusion coefficient of the bulk fluid. Therefore, a steeper slope in Fig. 6.4 at short diffusion times implies a smaller structure size. The steeper slope in $ADC_\perp$ implies that the structure size is smaller in the direction perpendicular to the fiber orientation. Values for the *S/V* for each direction were extracted from the curves shown in Fig. 6.4 using the value of D_0 for bulk water at 23°C. The resulting ratios were 2600 and 3700 cm^{-1} for the parallel and perpendicular directions, respectively.

6.1.4.3. Fast-Time Measurements of Tendon Water *ADC* with Loading and Unloading

In all freshly harvested tendons, the $ADC_\perp$ increased with load and decreased toward baseline upon unloading. Fig. 6.6 shows the response of the $ADC_\perp$ in a typical tendon. The increase in the $ADC_\perp$ following load exhibits the behavior of a typical step-response curve (Eq. [6.2], Fig. 6.5), while the reversal of the *ADC* back to baseline follows that of a relaxation curve following Eq. [6.3].

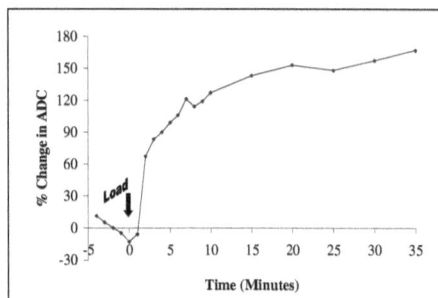

Fig. 6.5. Response of a fresh tendon to static 5-N loading. % *ADC* change plotted over the time course of the experiment. $ADC_\perp$ exhibits the behavior of a step response.

Fig. 6.7 shows the change in *ADC* with time upon loading and unloading. The rise-time behavior follows Eq. [6.5] while the fall-time behavior follows Eq. [6.6].

$$\%Change_{rise} = 52.6\left(1 - e^{-\frac{t}{3.9}}\right) \qquad [6.5]$$

$$\%Change_{fall} = 26.0e^{-\frac{t-5}{8.9}} + 0.1 \qquad [6.6]$$

The average rise time constant was 3.9 minutes while the average fall time constant was 8.9 minutes. Due to the short duration of the static load, the percentage change in *ADC* did not reach equilibrium and the unloading proceeded from a non-equilibrated value.

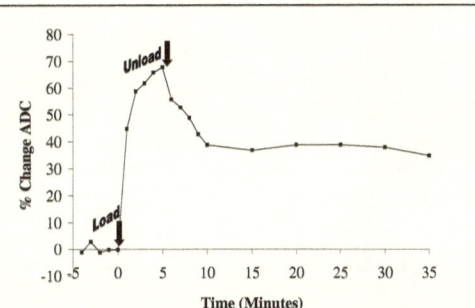

Fig. 6.6. Response of a fresh tendon to static 5-N loading and unloading. Percent *ADC* change is plotted over the time course of the experiment. This is one of the tendons where the *ADC* did not return to the baseline value.

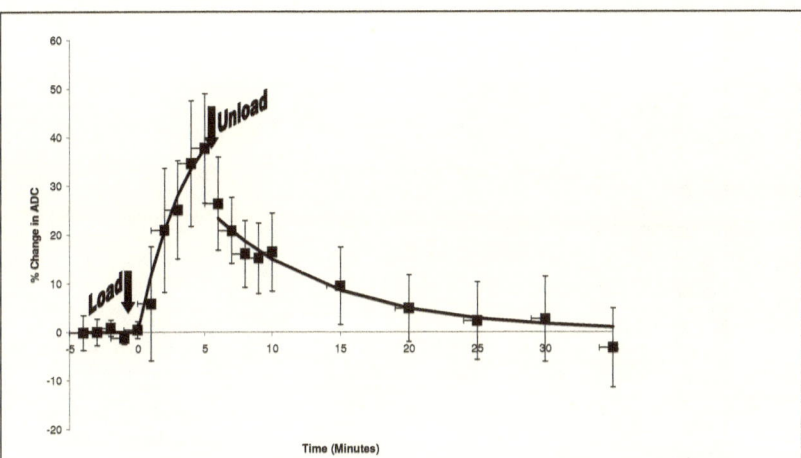

Fig. 6.7. Average tendon response curve to static 5-N loading and unloading (n=8). Average of % *ADC* change (±SEM) is plotted over the time course of the experiment. The *ADC* in all tendons increased subsequent to tensile loading and decreased with unloading. Data is for fresh tendons. Solid line represents the best fit.

TENDON

Several tendons did not recover to the baseline *ADC* value during the time course of the experiment.

6.1.4.4. Tendon Anisotropy and Effects of Storage Media

The tendons stored in PBS exhibited a gross swelling. This manifested itself in the *ADC* measurements of the tendons. For tendons stored in PBS, the $ADC_\| \approx 1.3 \times ADC_\perp$ at t_{diff} = 8.0 ms and $ADC_\| \approx 1.7 \times ADC_\perp$ at t_{diff} = 100.0 ms. Fresh tendons harvested immediately post mortem showed a greater degree of diffusional anisotropy (average

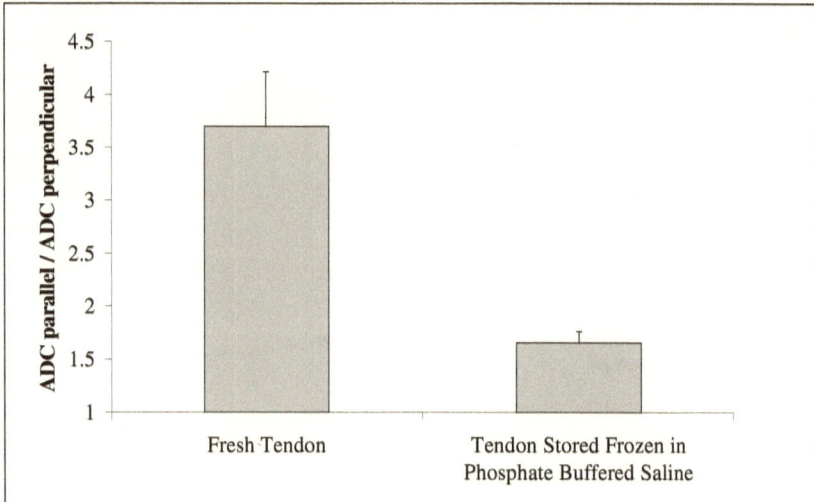

Fig. 6.8. Comparison of diffusional anisotropy for tendons stored frozen in phosphate-buffered saline and freshly harvested tendons. Diffusional anisotropy for t_{diff} = 80 ms. Chart shows $ADC_\| / ADC_\perp \pm$ SEM.

TENDON

$ADC_\perp = 1.42 \times 10^{-6}$ cm²/s, average $ADC_\parallel = 5.24 \times 10^{-6}$ cm²/s, $ADC_\parallel \approx 3.7 \times ADC_\perp$) at $t_{diff} = 80.0$ ms. Fig. 6.8 shows a comparison of diffusional anisotropy for tendons stored frozen in PBS and freshly harvested tendons. A larger change in $ADC_\perp$ is observed with a 5-N load (Table 6.1) for freshly-harvested tendons as compared to the previously stored-frozen tendons.

Table 6.1. Percent change in *ADC* for tendons stored-frozen in phosphate-buffered saline (PBS) and freshly-harvested tendons. Changes shown for 5-N load (± SEM).

	% change in *ADC*	Standard Error
Previously Stored-Frozen in PBS (n=10)	+5.1%	6.4%
Freshly-Harvested Tendons (n=8)	+37.7%	11.3%

6.1.4.5. Mechanical Response of Tendon to Loading

The stress-strain curve measured for tendons stored in PBS using the same loads as the NMR experiment are shown in Fig. 6.9. The stress-strain curves revealed that the loads used in this study caused the tissue to be on the non-linear toe region of the stress-strain curve. Maximum stresses, caused by the 10-N load, were on the order of 1.5 MPa.

In these experiments, the tendons were maintained in paraffin oil as static tensile loads were applied. Water was visibly extruded from the tendons during loading (Fig. 6.10), and water re-uptake was observed during periods of unloading.

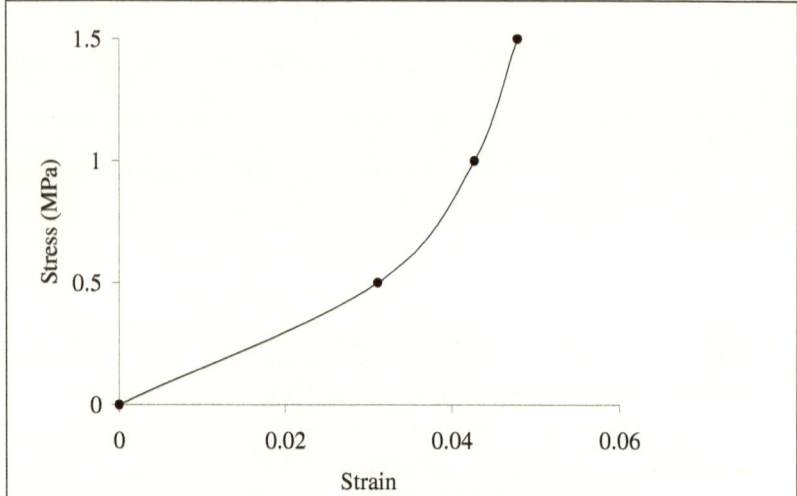

Fig. 6.9. Stress-strain curve for an Achilles tendon using loads up to 10 N. NMR experiments were performed in the nonlinear region of the curve. Analysis was done on previously frozen tissue stored in phosphate-buffered saline.

6.1.5. Discussion

The increase in *ADC* with load is consistent with the hypothesis that water is transported from a bound phase to a bulk phase. Since these are spectroscopic measurements, the measured *ADC* is a convolved average of the bound and bulk water in the tendon. It is believed that the uniaxial loading of the tendon causes an increase in the lateral pressure, resulting in the extrusion of water. Since the sample is immersed in a bath of paraffin oil, the extruded water is confined to the tendon surface and included in the *ADC* measurement. The collagen matrix within the tendon is structured in a helical or zigzag pattern (Diamant *et al.*, 1972; Kastelic *et al.*, 1978; Dale *et al.*, 1972; Comninou and

Yannas, 1976; Beskos and Jenkins, 1975; Millington *et al.*, 1971). It has been hypothesized that a water bridge is established between these helical structures *via* hydrogen bonds (Grigera and Berendson, 1979; Peto *et al.*, 1990). An uncrimping effect in these helical structures has been observed with load (Abrahams, 1967; Elliot, 1965). It has been suggested that the disruption of this coil geometry causes the bound water to be released into the bulk phase (Berendsen, 1962). Lim *et al.* (1971) have demonstrated structural changes at the molecular binding sites associated with mechanical deformation, lending credence to the explanation that bound water is being released. It might be expected from the report of Burstein *et al.* (1993) that the *ADC* of the water remaining inside the tendon would be reduced by loading. In that experiment, the water extruded from the cartilage was removed before *ADC*

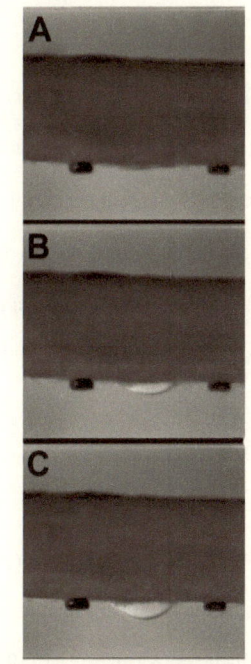

Fig. 6.10. Photograph showing the extrusion of bulk water from a rabbit Achilles tendon (stored in phosphate-buffered saline) under a 10-N load.
   A. At $t = 0$ min.
   B. At $t = 5$ min
        after application of load
   C. At $t = 13$ min
        after application of load
More typically, the extruded water formed a sheath around the tendon. This particular tendon was stored in phosphate-buffered saline before application of load and tested in paraffin oil.

measurements were performed. In the present experiment, signal changes from within the tendon would be outweighed by the net signal change due to the extrusion of water. The changes in *ADC* observed with load may also be attributed to damage caused by tensile loading. It has been shown that there are levels of strain beyond which the tendon exhibits plastic deformation (Partington and Wood, 1963; Rigby *et al.*, 1958; Abrahams, 1967). It is unlikely that such loads were used in our experiments. The stress-strain behavior of the tendon suggests that the stresses appear to be well below the levels at which the tendon would undergo failure.

The *ADC* differences between parallel and orthogonal directions can be accounted for in terms of tissue anisotropy. The *ADC* along a particular direction is decreased from the bulk value by structural impediments to diffusion along the measurement direction. The substructure of tendon consists of collagen molecules formed into fibrils of 20 – 40 nm diameter (Kastelic *et al.*, 1978). Groups of fibrils form fibers with diameters of 0.2 to 12 µm. These fibers are grouped into fascicles, which are sheathed to form the tendon. These fibrils are crimped to form a zigzag pattern with a period of ≈20-100 µm. Thus, the collagen component of the solid matrix offers only slight barriers to diffusion along the parallel direction while the diffusion orthogonal to the fiber orientation is more significantly restricted. We attribute the greater change in $ADC_\perp$ to the fact that diffusion along the orthogonal direction is more highly impeded and the shift of water from the bound to bulk phase would cause a greater increase in the *ADC*. The fresh tendon exhibited a higher degree of diffusional anisotropy ($ADC_\parallel / ADC_\perp$) than the swollen tendon. This suggests that the distances between the elements of the collagen matrix are increased due to tendon swelling. Thus, when a load is applied, the tendon water is squeezed out of the tendon (Haut and Powlison, 1990). Since increased restriction effects are expected for $ADC_\perp$, a larger change in ADC would be expected with load for fresh tendons as compared to swollen tendons. The greater change in ADC associated with fresh tendons is consistent with the idea that the water molecules are more restricted in fresh tendons, since the shift of water into the bulk phase would result in a larger change in the measured ADC.

From the initial slope in Fig. 6.3, the surface area to volume ratio (S/V) can be found assuming a known value of the bulk phase diffusion coefficient, D_0. To show that the data in Fig. 6.4 implies a difference in restrictions in the two directions, we have calculated the S/V in both cases using a D_0 of bulk water at 23°C and found that the S/V parallel to the fiber orientation is roughly 70% of S/V in the orthogonal direction. This result confirms the finding of a difference in characteristic size in the two orthogonal directions. The use of D_0 for bulk water is justified since only a comparative difference is being made. The actual value of D_0 would be necessary for absolute values of S/V to be determined.

The finding of anisotropic diffusion of water in rabbit Achilles tendon in the present experiment conflicts with the finding of diffusional isotropy by Henkelman *et al.* (1994) in diffusion experiments on bovine Achilles tendon. One possible explanation is that the

TENDON *147*

bovine Achilles tendon could exhibit a larger separation of the collagen fibers and a shorter crimping period as compared to the rabbit tendon. Since the size of these substructures is dependent on the species as well as tissue types (Kastelic *et al.*, 1978; Evans and Barbenel, 1974), it is possible that the diffusional characteristics of tendon in these two species is different. Due to the short diffusion time used by Henkelman *et al.* (19.0 ms), it is possible that the restriction effects are minimal for the bovine Achilles tendon. Another possible explanation is that in order to measure the $ADC_\perp$, Henkelman *et al.* had to rotate the sample in order to change the gradient direction in their measurements. The strong orientation of the sample fibrils may cause an angular dependence in susceptibility patterns and result in a similar *ADC* measured in both the perpendicular and parallel directions. Because the gradient direction was varied and the sample orientation kept constant for the present experiment, this orientation dependent anisotropy is not expected to affect our results. In addition, for the shortest diffusion times, we used an anti-symmetric bipolar gradient pair for diffusion sensitization. This has been shown to reduce the effects of sample susceptibility on the measured *ADC* (Trudeau *et al.*, 1995).

The *ADC*-response curve to the static uniaxial load looks like the expected deformation behavior of a Voight solid subjected to a sudden load. During the loading and unloading periods, the behavior of the *ADC* was consistent with a response to a step function (i.e. sudden loads and releases). Although the magnitude of the $ADC_\perp$ may have been different for each of the tendons, the behavior of the rise and fall times was consistent from tendon to tendon. Since this behavior is consistent with models describing the mechanical behavior of viscoelastic solids (Ault and Hoffman, 1992a, 1992b), it is suggested that the *ADC* increase is integrally related to the structural changes associated with mechanical loading. It has been shown that the mechanical behavior follows a time course where a finite amount of time is required before the tendon reaches equilibrium after loading (Abrahams, 1967; Hannafin and Arnoczky, 1994). The time constant for the change in percent $ADC_\perp$ with loading is much faster than that of unloading. This suggests that the mechanism behind water extrusion might be different than that of water uptake by the tendon.

One of the concerns is that the *ADC* measurements are contaminated by a "creep" effect. Since *ADC* measurements are sensitive to motion, any stretching of the material would manifest itself as an increase in *ADC* (Reese *et al.*, 1996). In the current experiment, since the load is immediately followed by the *ADC* measurements, we cannot be certain whether material stretching contributes significantly to the increased *ADC*. However, since the load is applied axially while the measurements are made transversely, the effect of material creep should be minimal in the measurements.

For this set of experiments, only the *ADC* orthogonal to the fiber orientation was measured. It has been speculated that the diffusion along the parallel axis of the fiber is relatively unimpeded compared to the diffusion transverse to the fiber. Hence, the changes in *ADC* parallel to the fiber orientation would be relatively small compared to the changes in *ADC* in the perpendicular direction.

Another point to note is that the *ADC* of several tendons never recovered after the load. This suggests that the process of extrusion is either not completely reversible in an excised tendon, or that the 5-N load irreparably damages the tendon in some way. The measurement of tendon water *ADC* over a fast time course suggests that the loading and unloading characteristics of the tendon follow a typical step response of a Voight solid. Since the *ADC* increase is consistent with the hypothesis that water is being extruded with loading and the process is reversible to some extent, it is believed that tendons are biphasic materials whose properties are determined by mixture theory (Mow *et al.*, 1984).

Chapter 7

Summary

In the study involving radiotherapy of murine tumors, the scientific question being addressed is that of tumor oxygenation, its relation to radiation therapy, and the ability to noninvasively measure the effects of radiation on tumor oxygenation. The importance of such a study is that the behavior of tumors in response to radiation will allow for strategic use of such therapies in the treatment of human cancers. During the course of these experiments, we found that there was a discrepancy between the spectroscopic and imaging data. My contribution to this study was in reconciling the differences between the spectroscopic and imaging data, allowing for a direct comparison between these two different modalities. This work was presented at the IEEE conference where it took first place. This study was a collaboration between Drs. Sotak, Dardzinski, and myself.

The study involving the correlation between $D(t)$ of tumor ^1H_2O and pO_2 derived its significance from the fact that it challenged an existing theory. While it is difficult to challenge peer reviewed published work, this study showed on a larger scale that theories evolve as information accumulates. Although this work was not the effort of one individual, the insight of the principal investigator (Dr. Karl Helmer) and his direction in the study proved to be rewarding. This set an example of the need for good direction and

SUMMARY

clear vision in science. I was the second author on this paper, and my contribution was in the animal care, data acquisition, and part of the data processing.

On the surface, the study involving yeast cell suspensions appears to be a validation of a method. Indeed, this is true, but the importance of these experiments lie in the application of the method. Due to the complexity of interpreting data obtained from animal experiments, it often becomes necessary to take a step back and find an analogous model to answer questions arising from the animal experiments. Specifically, the question that vexed us was the diffusion behavior of water in a two-compartment system. During cerebral ischemia, the brain water apparent diffusion coefficient (ADC) decreases, and there are many plausible explanations behind this phenomena. Others have done studies involving intra- and extracellular metabolites to infer the behavior of the water diffusion coefficient from the behavior of these metabolites, but it is not evident *a priori* that the diffusion characteristics of these metabolites track those of water. Not only so, but metabolites that are constrained to only one compartment cannot accurately represent water, which can move between the two compartments. This study was a stepping stone to answering the perplexing question of what is really happening to the brain water at the onset of ischemia. I was responsible for working out the specific theory behind the experiments, designing the experiments, writing the pulse sequences, acquiring the data, and analyzing the data. Although this work was a collaborative effort of many, I was the one who set the direction of these experiments and analyzed the data to answer the underlying question of compartmental ^1H_2O ADC behavior. The future direction of this

SUMMARY

particular study would be to further characterize the water behavior in yeast by measuring the time-dependence of the *ADC* to extract structural parameters from the cellular compartments. Once this is done, *in vivo* measurements of compartmental water *ADC* in the rat brain under different pathological conditions would need to be done.

The rabbit Achilles tendon experiment showed the true multi-faceted nature of some NMR studies. The characterization of a system that is not well known is not a trivial task, and in this case, it was especially necessary to collaborate with scientists from multi-disciplinary backgrounds. Since the mechanical response to a biological system was under observation using NMR, mechanical engineers, physiologists, biomedical engineers, and NMR scientists were all involved in every step of these experiments. Although many people collaborated in this study, I was the one who set the direction for these studies. Since the system was not well characterized, there was a need to narrow the overall scope of this study and specify which questions needed to be addressed. I was responsible for forming the main scientific questions that needed to be addressed and working out the specific experiments necessary to characterize the system. I also wrote the pulse sequences necessary for the data acquisition and wrote the code to analyze all of the data. Since this study involving the rabbit Achilles tendon is fairly new, there is much that can be done with this and also many different directions in which this study can go. One point of interest is that the behavior of visco-elastic solids is different for dynamic loading conditions that for static loading conditions. The study of the water behavior to dynamic loading would be an interesting avenue to explore. Furthermore, due to the highly structured fiber orientation of the tendon, it would be interesting to know if there are any directional susceptibility artifacts that manifest themselves as a function of the tendon's orientation in the static magnetic field.

Bibliography

Abragam A, Principles of Nuclear Magnetism, Oxford University Press, London, 1961.

Abrahams M, Mechanical Behavior in Tendon *In vitro*: A Preliminary Report, *Med. Biol. Eng.*, **5**, 433-443, 1967.

Andrasko J, Measurement of Membrane Permeability to Slowly Penetrating Molecules by a Pulse Gradient NMR Method, *J. Magn. Reson.*, **21**, 479-484, 1976.

Andrasko J, Water Diffusion Permeability of Human Erythrocytes Studied by a Pulsed Gradient NMR Technique, *Biochim. Biophys. Acta*, **428**, 304-311, 1976.

Ault HK, Hoffman AH, A Composite Micromechanical Model for Connective Tissues: Part I – Theory, *J. Biomech. Eng.*, **114**, 137-141, 1992.

Ault HK, Hoffman AH, A Composite Micromechanical Model for Connective Tissues: Part II – Application to Rat Tail Tendon and Joint Capsule, *J. Biomech. Eng.*, **114**, 142-146, 1992.

Bacic G, Liu KJ, O'Hara JA, Harris RD, Szybinski K, Swartz HM, Oxygen Tension in a Murine Tumor: a Combined EPR and MRI Study, *Magn. Reson. Med.*, **30**, 568-572, 1993.

Baldwin NM, Ng TC, pO_2 Maps of Murine Tumors as Determined by ^{19}F MRI, *In. Proc. SMRM, 11th Annual Meeting*, Berlin, p. 3501, 1992.

Barker BR, Mason RP, Bansal N, Peshock RM, Oxygen Tension Mapping by ^{19}F Echo Planar NMR Imaging of Sequestered Perfluorocarbon, *J. Magn. Reson. Imaging*, **4**, 595-602, 1994.

Barker BR, Mason RP, Peshock RM, Echo Planar Imaging of Perfluorocarbons, *Magn. Reson. Imaging*, **11**, 1165-1173, 1993.

Beaulieu C, Allen PS, Water Diffusion in the Giant Axon of the Squid: Implications for Diffusion-Weighted MR of the Nervous System, *Magn. Reson. Med.*, **32**, 579-583, 1994.

Berendsen HJC, Nuclear Magnetic Resonance Study of Collagen Hydration, *J. Chem. Phys.*, **36(12)**, 3297-3305, 1962.

Berkowitz BA, Wilson CA, Hatchell DL, London RE, Quantitative Determination of the Partial Oxygen Pressure in the Vitrectomized Rabbit Eye *In Vivo* Using ^{19}F NMR, *Magn. Reson. Med.*, **21**, 233-241, 1991.

Beskos DE, Jenkins JT, A Mechanical Model for Mammalian Tendon, *J. Appl. Mech.*, **42**, 755-758, 1975.

Betsch DF, Baer E, Structure and Mechanical Properties of Rat Tail Tendon, *Biorheology*, **17**, 83-94, 1980.

Bhujwalla ZM, Glockner JF, Artemov D, Glickson JD, Magnetic Resonance Imaging of Tumor Vascular Volume and Permeability of Metastatic Human Breast Cancer and Non-Metastatic Rodent Fibrosarcoma, *In. Proc. ISMRM, 4th Annual Meeting*, p. 348, New York, 1996.

Bloch F, Hansen WW, Packard M, Nuclear Induction Experiment, *Physiol. Rev.*, **70**, 474-485, 1946.

Bloch F, Nuclear Induction, *Physiol. Rev.*, **70**, 460-474, 1946.

Boucher Y, Baxter LT, Jain RK, Interstitial Pressure Gradients in Tissue-Isolated and Subcutaneous Tumors: Implications for Therapy, *Cancer Res.*, **50**, 4478-4484, 1990.

Boucher Y, Jain RK, Microvascular Pressure is the Principle Driving Force for Interstitial Hypertension in Solid Tumors: Implications for Vascular Collapse, *Cancer Res.*, **52**, 5110-5114, 1992.

Brown JM, Evidence for Acutely Hypoxic Cells in Mouse Tumors and a Possible Mechanism of Reoxygenation, *Br. J. Radiol.*, **52**, 650-656, 1979.

Burstein D, Gray ML, Hartman AL, Gipe R, Foy BD, Diffusion of Small Solutes in Cartilage as Measured by Nuclear Magnetic Resonance (NMR) Spectroscopy and Imaging, *J. Orthop. Res.*, **11**, 465-478, 1993.

Callaghan PT, Principles of Nuclear Magnetic Resonance Microscopy, Oxford Science Publications, New York, New York, 1991.

Ceckler TL, Gibson SL, Hilf R, Bryant RG, *In Situ* Assessment of Tumor Vascularity Using Fluorine NMR Imaging, *Magn. Reson. Med.*, **13**, 416-433, 1990.

Chaplin DJ, Olive PL, Durand RE, Intermittent Blood Flow in a Murine Tumor: Radiobiological Effects, *Cancer Res.*, **47**, 597-601, 1987.

Chimich D, Shrive N, Frank C, Marchuk L, Bray R, Water Content Alters Viscoelastic Behaviour of the Normal Adolescent Rabbit Medial Collateral Ligament, *J. Biomech*, **25**, 831-837, 1992.

Clark Jr. LC, Thomas SR, Pratt RG, Hoffman RE, Busse LJ, Samaratunga RC, Ackerman JL, NMR Determination of Liver pO_2 *In Vivo* Using Perfluorocarbon Emulsions, *In. Proc. SMRM, 4th Annual Meeting*, p. 40-41, London, England, 1985.

Clark LC, Ackerman JL, Thomas SR, Millard RW, Hoffman RE, Pratt RG, Ragel-Cole H, Kinsey RA, Janakiraman R, Perfluorinated Organic Liquids and Emulsions as Biocompatible NMR Imaging Agents for ^{19}F and Dissolved Oxygen, *Adv. Exp. Med. Biol.*, **180**, 835-845, 1984.

Comninou M, Yannas IV, Dependence of Stress-Strain Nonlinearity of Connective Tissues on the Geometry of Collagen Fibers, *J. Biomech.*, **9**, 427-433, 1976.

Conlon T, Outhred R, Water Diffusion Permeability of Erythrocytes Using an NMR Technique, *Biochim. Biophys. Acta*, **288**, 354-361, 1972.

Dale WC, Baer E, Keller A, Kohn RR, On the Ultrastructure of Mammalian Tendon, *Experientia* **28**, 1293-1295, 1972.

Dardzinski BJ, Palyka I, Reith W, Springer CS Jr., Sotak CH, Monitoring of Hyperosmotic Blood-Brain-Barrier Opening to Contrast Agent in Rats by Relaxographic Imaging, *In. Proc. ISMRM, 2nd Annual Meeting*, p. 1382, San Francisco, CA, 1994.

Dardzinski BJ, Sotak CH, Evaluating Changes in Murine Tumor Oxygenation in Response to Nicotinamide by Using ^{19}F Echo-Planar Imaging and Spectroscopy, *J. Magn. Reson. Imaging*, **4(P)**, 55, 1994.

Dardzinski BJ, Sotak CH, *In Vivo* Oxygenation Mapping with Inversion-Recovery Echo-Planar Imaging and ^{19}F Relaxometry of Perfluorocarbon Emulsions., *J. Magn. Reson. Imaging*, **3(P)**, 157-158, 1993.

Dardzinski BJ, Sotak CH, *In Vivo* Tumor Oxygenation Mapping Using Inversion-Recovery Echo-Planar Imaging and ^{19}F Relaxometry of Perfluorocarbon Emulsions, *In. Proc. SMRM, 12th Annual Meeting*, p. 405, New York, New York, August, 1993.

Dardzinski BJ, Sotak CH, *In Vivo* Tumor Oxygenation Using Inversion-Recovery Echo-Planar Imaging and ^{19}F Relaxometry of Perfluorocarbon Emulsions, *In. Proc. SMRM, 12th Annual Meeting*, p. 403, New York, New York, August, 1993.

Dardzinski BJ, Sotak CH, Isabelle A, Murray J, Stojadinovic P, Mapping the Tissue Oxygen Tension in RIF-1 Tumors Following Radiation Therapy Using Inversion-Recovery Echo-Planar Imaging and ^{19}F Relaxometry of Perfluoro-15-Crown-5-Ether, *In. Proc. ISMRM, 2nd Annual Meeting*, p. 427, San Francisco, CA, August, 1994.

Dardzinski BJ, Sotak CH, Rapid Tissue Oxygen Tension Mapping Using ^{19}F Inversion-Recovery Echo-Planar Imaging of Perfluoro-15-Crown-5-Ether, *Magn. Reson. Med.*, **32**, 88-97, 1994.

Diamant J, Keller A, Baer E, Litt M, Arridge RGC, Collagen: Ultrastructure and its Relation to Mechanical Properties as a Function of Aging, *Proc. R. Soc. London*, **180**, 293-315, 1972.

Dunn JF, Ding S, O'Hara JA, Liu KJ, Rhodes E, Weaver JB, Swartz HM, The Apparent Diffusion Constant Measured by MRI Correlates with pO_2 in a RIF-1 Tumor, *Magn. Reson. Med.*, **34**, 515-519, 1995.

Duquette JJ, Grigg P, Hoffman AH, The Effect of Diabetes on the Viscoelastic Properties of Rat Knee Ligaments, *J. Biomech. Eng.*, **118**, 557-564, 1996.

Eidelberg D, Johnson G, Barnes D, Tofts PS, Delpy D, Plummer D, McDonald WI, ^{19}F NMR Imaging of Blood Oxygenation in the Brain, *Magn. Reson. Med..*, **6**, 344-352, 1988.

Eidelberg D, Johnson G, Tofts PS, Dobbin J, Crockard HA, Plummer D, ^{19}F Imaging of Cerebral Blood Oxygenation in Experimental Middle Cerebral Artery Occlusion: Preliminary Results, *J. Cereb. Blood Flow Metab.*, **8**, 276-281, 1988.

Eldon HR, Hydration of Connective Tissue and Tendon Elasticity, *Biochim. Biophys. Acta*, **79**, 592-599, 1964.

Elliot DH, Structure and Function of Mammalian Tendon, *Biol. Rev. Camb. Philos. Soc.*, **40**, 392-421, 1965.

Evans JH, Barbenel JC, Structure and Mechanical Properties of Tendon Related to Function, *Equine Vet. J.*, **7**, i-viii, 1974.

Fishman JE, Joseph PM, Carvlin MJ, Saadi-Elmandjra M, Mukherji B, Sloviter HA, *In Vivo* Measurements of Vascular Oxygen Tension in Tumors Using MRI of a Fluorinated Blood Substitute, *Invest. Radiol.*, **24**, 65-71, 1987.

Fishman JE, Joseph PM, Floyd TF, Mukherji B, Sloviter HA, Oxygen-Sensitive ^{19}F NMR Imaging of the Vascular System *In Vivo*, *Magn. Reson. Imaging*, **5**, 279-285, 1987.

Fox SI, Human Physiology, 4th Edition, Wm. C. Brown Publishers, Dubuque, Iowa, 1993.

Frisen M, Magi M, Sonnerup L, Viidik A, Rheological Analysis of Soft Collagenous Tissue. Part II: Experimental Evaluation and Verifications, *J. Biomech.*, **2**, 21-28, 1969.

Fung YC, Biomechanics: Mechanical Properties of Living Tissues, Second Edition, New York, Sprinter-Verlag, 1993.

BIBLIOGRAPHY

Galante JO, Tensile Properties of the Human Lumbar Annulus Fibrosus, *Acta Ortho. Scand. Suppl.*, **100**, 1-91, 1967.

Gemmell SJ, Investigation of Viscoelastic Properties of Tendon Using Nuclear Magnetic Resonance, M. S. Thesis, WPI, 1996.

Gray LH, Conger AD, Ebert M, Hornsey S, Scott OCA, Concentration of Oxygen Dissolved in Tissues at Time of Irradiation as a Factor in Radiotherapy, *Br. J. Radiol.*, **26**, 638-648, 1953.

Grigera JR, Berendson HJC, The Molecular Details of Collagen Hydration, *Biopolymers*, **18**, 47-57, 1979.

Hahn EL, Spin Echoes, *Phys. Rev.*, **80**, 580-594, 1950.

Halliday D, Resnick R, Fundamentals of Physics, Third Edition, John Wiley & Sons, 1988.

Han SS, Dardzinski BJ, Sotak CH, Reconciling Differences between Tumor Oxygenation Measurements Performed Using ^{19}F NMR Spectroscopy and Imaging of Perfluorocarbon Emulsions, *In.. Proc. IEEE NE Bioeng. Conf.*, 23rd *Annual Meeting*, p. 63, Durham, NH, May, 1997.

Han SS, Vétek G, Silva MD, Springer CS, Sotak CH, Deconvolution of Restriction Effects on Compartmental Diffusion Using Combined Relaxography and Diffusion Measurements, *In. Proc. ENC*, 39th *Annual Meeting*, p. 264, Pacific Grove, CA, 1998.

Han SS, Vétek G, Springer CS, Sotak CH, Apparent Diffusion Coefficients of Intra- and Extracellular Water in Yeast Suspensions Measured by Combined Diffusion and Relaxography, *In. Proc. ISMRM*, 6th *Annual Meeting*, p. 535, Sydney, Australia, 1998.

Hannafin JA, Arnoczky SP, Effect of Cyclic and Static Tensile Loading on Water Content and Solute Diffusion in Canine Flexor Tendons, an *In vitro* Study, *J. Orthop. Res.*, **12**, 350-356, 1994.

Haus JW, Kehr KW, Diffusion in Regular and Disordered Lattices, *Phys. Rep.*, **150**, 263-406, 1987.

Haut RC, Powlison AC, The Effects of Test Environment and Cyclic Stretching on the Failure Properties of Human Patellar Tendons, *J. Orthop. Res.*, **8**, 532-540, 1990.

Hees PS, Sotak CH, Assessment of Changes in Murine Tumor Oxygenation in Response to Nicotinamide Using ^{19}F NMR Relaxometry of a Perfluorocarbon Emulsion, *Magn. Reson. Med.*, **29**, 303-310, 1993.

Helmer KG, Dardzinski BJ, Sotak CH, The Application of Porous Media Theory to the Investigation of Time-Dependent Diffusion in *In vivo* Systems, *NMR Biomed.*, **8**, 297-306, 1995.

Henkelman RM, Stanisz GH, Kim JK, Bronskill MJ, Anisotropy of NMR Properties of Tissues, *Magn. Reson. Med.*, **32**, 592-599, 1994.

Herbst MD, Goldstein JH, A Review of Water Diffusion Measurement by NMR in Human Red Blood Cells, *Am. J. Physiol.*, **256**, C1097-C1104, 1989.

Holland SK, Kennan RP, Schaub MM, D'Angelo MJ, Gore JC, Imaging Oxygen Tension in Liver and Spleen by ^{19}F NMR, *Magn. Reson. Med.*, **29**, 446-458, 1993.

Hong X, Dixon WT, Measuring Diffusion in Inhomogeneous Systems in Imaging Mode Using Antisymmetric Sensitizing Gradients, *J. Magn. Reson.*, **99**, 561-570, 1992.

Hooley CJ, McCrum NG, Viscoelastic Creep of Collagenous Tissue, *J. Biomech.*, **9**, 175-184, 1976.

Hore PJ, Solvent Suppression in Fourier Transform Nuclear Magnetic Resonance, *J. Magn. Reson.*, **55**, 283-300, 1983.

Jain RK, Transport of Molecules in the Tumor Interstitium: A Review, *Cancer Res.*, **47**, 3039-3051, 1987.

Jap BK, Li H, Structure of the Osmo-Regulated H₂O-Channel, AQP-CHIP, in Projection at 3.5 Å Resolution, *J. Mol. Biol.*, **251**, 413-420, 1995.

Johnson LM, Plona TJ, Scala C, Pasierb F, Kojima H, Tortuosity and Acoustic Slow Waves, *Phys. Rev. Lett.*, **49**, 1840-1844, 1982.

Joseph PM, A Spin Echo Chemical Shift MR Imaging Technique, *J. Comput. Assist. Tomogr.*, **9**, 651-658, 1985.

Joseph PM, Yuasa Y, Kundel HL, Mukherji B, Sloviter HA, Magnetic Resonance Imaging of Fluorine in Rats Infused with Artificial Blood, *Invest. Radiol.*, **20**, 504-509, 1985.

Kallman RF, The Phenomenon of Reoxygenation and It's Implications for Fractionated Radiotherapy, *Radiology*, **105**, 135-142, 1972.

Kärger J, Pfeiffer H, Heink W, Principle and Application of Self-Diffusion Measurements by Nuclear Magnetic Resonance, *Adv. Magn. Reson.*, **12**, 1-89, 1988.

Kastelic J, Baer E, A Structural Mechanic Model for Tendon Crimping, *J. Biomech.*, **13**, 887-893, 1980.

Kastelic J, Galeski A, Baer E, The Multicomposite Structure of Tendon, *J. Conn. Tiss. Res.*, **6**, 11-23, 1978.

Kleinberg RL, Kenyon WE, Mitra PP, Mechanism of NMR Relaxation in Fluids in Rock, *J. Magn. Reson. A*, **108**, 206-214, 1994.

Knight RA, Ordidge RJ, Helpern JA, Chopp M, Rodolosi LC, Peck D, Temporal Evolution of Ischemic Damage in Rat Brain Measured by Proton Nuclear Magnetic Resonance Imaging, *Stroke*, **22**, 802-808, 1991.

Kong CF, Holloway GM, Parhami P, Fung BM, Carbon-13 and Fluorine-19 Nuclear Magnetic Resonance of Perfluoro Chemical Emulsions, *J. Phys. Chem.*, **88**, 6308-6311, 1984.

Krishnan CE, Krishnan L, Jewell B, Bhatia P, Jewell WR, Dose-Dependent Radiation Effect on Microvasculature and Repair, *J. Nat. Cancer Inst.*, **79**, 1321-1325, 1987.

Kruv JA, Inch WR, McCredie JA, Blood Flow and Oxygenation of Tumors in Mice. I. Effects of Breathing Gases Containing Carbon Dioxide at Atmospheric Pressure, *Cancer*, **20**, 51-70, 1967.

Kwan MK, Lin T H-C, Woo S L-Y, On the Viscoelastic Properties of the Anteromedial Bindle of the Anterior Cruciate Ligament, *J. Biomech.*, **26(4/5)**, 447-452, 1993.

Labadie C, Lee J-H, Vetek G, Springer CS Jr., Relaxographic Imaging, *J. Magn. Reson. B*, **105**, 99-112, 1994.

Laird NM, Ware JH, Random-Effects Model for Longitudinal Data, *Biometrics*, **38**, 963-974, 1982.

Lanir Y, Structure-Strength Relations in Mammalian Tendon, *Biophys. J.*, **24**, 541-554, 1978.

Latour LL, Mitra PP, Kleinberg RL, Sotak CH, Time-Dependent Diffusion Coefficient of Fluids in Porous Media as a Probe of Surface-to-Volume Ratio, *J. Magn. Reson. A*, **101**, 342-346, 1993.

Latour LL, Svoboda K, Mitra PP, Sotak CH, Observation of Changes in the Apparent Diffusion Coefficient of Water with Changes in Membrane Permeability in Red Blood Cells, *In. Proc. SMRM, 11th Annual Meeting*, San Francisco, CA, 1993; Also published in *J. Magn. Reson. Imaging*, **3(P)**, 90, 1993.

Latour LL, Svoboda K, Mitra PP, Sotak CH, Time-Dependent Diffusion of Water in a Biological Model System, *Proc. Natl. Acad. Sci. USA*, **91**, 1229-1233, 1994.

LeBihan D, Basser PJ, Diffusion and Perfusion Magnetic Resonance Imaging, Raven Press, New York, 1995.

LeBihan D, Molecular Diffusion Nuclear Magnetic Resonance Imaging, *Magn. Reson. Q.*, **7**, 1-30, 1991.

LeBihan D, Turner R, Moonen CTW, Pekar J, Imaging of Diffusion and Microcirculation with Gradient Sensitization: Design, Strategy, and Significance, *J. Magn. Reson. Imgaging*, **1**, 7-28, 1991.

Lim JJ, Shamos MH, An Investigation of the Bound Water in Tendon by Dielectric Measurement, *Biophys. J.*, **11**, 648-663, 1971.

Longmaid HE, Adams DF, Neirinckx RD, Harrison CG, Brunner P, Seltzer SE, Davis MA, Neuringer L, Geyer RP, In Vivo ^{19}F Imaging of Liver, Tumor, and Abscess in Rats: Preliminary Results, *Invest. Radiol.*, **20**, 141-145, 1985.

Mansfield P, Grannell PK, "Diffraction" and Microscopy in Solids and Liquids by NMR, *Phys. Rev. B*, **12**, 3618, 1975.

Mansfield P, Multi-Planar Image Formation Using NMR Spin Echoes, *J. Phys. C. Solid State Phys.*, **10**, L55-58, 1977.

Mason RP, Nunnally RL, Antich PP, Tissue Oxygenation: A Novel Determination Using ^{19}F Surface Coil NMR Spectroscopy of Sequestered Perfluorocarbon Emulsion, *Magn. Reson. Med.*, **18**, 71-79, 1991.

Mason RP, Shukla H, Antich PP, Oxygent^{TM}: A Novel Probe of Tissue Oxygen Tension, *Biomater. Artif. Cells Immobilization Biotechnol.*, **20**, 929,932, 1992.

McFarland E, Koutcher HA, Rosen BR, Teicher B, Brady TJ, In Vivo ^{19}F NMR Imaging, *J. Comput. Assist. Tomogr.*, **9**, 8-15, 1985.

Millington PF, Gibson TG, Evans JH, Barbenel TC, Structural and Mechanical Aspects of Connective Tissue, *Adv. Biomed. Eng.*, **1**, 189-248, 1971.

Mitra PP, Sen PN, Schwartz LM, Le Doussal P, Diffusion Propagator as a Probe of the Structure of Porous Media, *Phys. Rev. Lett.*, **68**, 3555-3558, 1992.

Mitra PP, Sen PN, Schwartz LM, Short-Time Behavior of the Diffusion Coefficient as a Geometrical Probe of Porous Media, *Phys. Rev. B*, **47**, 8565-8574, 1993.

Morris P, Nuclear Magnetic Resonance Imaging in Medicine and Biology, Clarendon Press, Oxford, 1986.

Moseley ME, Cohen Y, Mintorovictch J, Kucharczyk J, Weinstein PR, Early Detection of Regional Cerebral Ischemia in Cats: Comparison of Diffusion- and T_2-Weighted MRI and Spectroscopy, *Magn. Reson. Med.*, **14**, 330-346, 1990.

Mow VC, Holmes MH, Lai WM, Fluid Transport and Mechanical Properties of Articular Cartilage, *J. Biomech.*, **17**, 377-394, 1984.

Nicholson C, Phillips JM, Gardner-Medwin AR, Diffusion from an Iontophoretic Point Source in the Brain: Role of Tortuosity and Volume Fraction, *Brain Res.*, **169**, 580-584, 1979.

Nicholson C, Phillips JM, Ion Diffusion Modified by Tortuosity and Volume Fraction in the Extracellular Microenvironment of the Rat Cerebellum, *J. Physiol.*, **321**, 225-257, 1981.

Nicholson C, Rice ME, Diffusion of Ions and Transmitters in the Brain Cell Microenvironment, Volume Transmissions in the Brain: Novel Mechanism for Neural Transmission, Fluxe K, Agnati LF, 279-294, Raven Press, New York, 1991.

Niendorf T, Dijkhuizen RM, Norris DG, van Lookeren Campagne M, Nicolay K, Biexponential Diffusion Attenuation in Various States of Brain Tissue: Implications for Diffusion-Weighted Imaging, *Magn. Reson. Med.*, **36**, 847-857, 1996.

Niendorf T, Norris DG, Dreher W, Leibfritz D, The Significance of Biexponential Diffusion Attenuation Curves in Brain Tissue, *In. Proc. ISMRM, 2nd Annual Meeting*, p. 139, San Francisco, 1994.

Noggle JH, Schirmer RE, The Nuclear Overhauser Effect: Chemical Applications, Academic, New York, 1971.

Norris DG, Niendorf T, Interpretation of DW-NMR Data: Dependence on Experimental Conditions, *NMR Biomed.*, **8**, 280-288, 1995.

Olive PL, Vikse C, Martin AA, Trotter J, Measurement of Oxygen Diffusion Distance in Tumor Cubes Using a Fluorescent Hypoxia Probe, *Int. J. Radiat. Oncol. Biol. Phys.*, **22**, 397-402, 1992.

Pályka I, Dardzinski BJ, Sotak CH, Springer CS Jr., The Implementation of Longitudinal Relaxographic Imaging with an EPI-Based Pulse Sequence and at Clinical Field Strength, *In. Proc. ISMRM, 2nd Annual Meeting*, p. 1394, San Francisco, CA 1994.

Parhami P, Fung BM, Fluorine-19 Relaxation Study of Perfluoro Chemicals as Oxygen Carriers, *J. Phys. Chem.*, **87**, 1928-1931, 1983.

Partington FR, Wood GC, The Role of Non-Collagen Components in the Mechanical Behavior of Tendon Fibres, *Biochim. Biophys. Acta.*, **69**, 485-495, 1963.

Peebles PZ, Probability, Random Variables, and Random Signal Principles, Second Edition, McGraw-Hill Book Company, 1987.

Peto S, Gillis P, Henri VP, Structure and Dynamics of Water in Tendon from NMR Relaxation Measurements, *Biophys. J.*, **57**, 71-84, 1990.

Pilatus U, Shim H, Artemov D, Davis D, van Zijl PCM, Glickson JD, Intracellular Volume and Apparent Diffusion Constants of Perfused Cell Cultures, As Measured by NMR, *Magn. Reson. Med.*, **37**, 825-832, 1997.

Press WH, Flannery BP, Teukolsky SA, Vetterling WI, Numerical Recipes in C, The Art of Scientific Computing, 683-688, Cambridge University Press, Cambridge, 1988.

Price WS, Pulsed-Field Gradient Nuclear Magnetic Resonance as a Tool for Studying Translational Diffusion: Part 1. Basic Theory, *Concepts Magn. Reson.*, 299-335, 1997.

Provencher SW, A Constrained Regularization Method for Inverting Data Represented by Linear Algebraic or Integral Equations, *Comput. Phys. Commun.*, **27**, 213-227, 1982.

Provencher SW, CONTIN: A General Purpose Constrained Regularization Program for Inverting Noisy Linear Algebraic and Integral Equations, *Comput. Phys. Commun.*, **27**, 229-242, 1982.

Provencher SW, Dovi VG, Direct Analysis of Continuous Relaxation Spectra, *J. Bioch. Biophys. Meth.*, **1**, 313-318, 1979.

Ratner AV, Muller HH, Simpson BB, Johnson DE, Hurd RE, Sotak CH, Young SW, Detection of Tumors with ^{19}F Magnetic Resonance Imaging, *Invest. Radiol.*, **23**, 361-364, 1988.

Ratner AV, Sotak CH, Muller H, Hurd R, Young SW, ^{19}F Magnetic Resonance Imaging of the Reticuloendothelial System, *Magn. Reson. Med.*, **5**, 548-554, 1987.

Reese TG, Wedeen VJ, Weisskoff RM, Measuring Diffusion in the Presence of Material Strain, *J. Magn. Reson.*, **112**, 253-258, 1996.

Reid RS, Koch CJ, Castro ME, Lunt JA, Treiber EO, Biosvert DJP, Allen PS, The Influence of Oxygenation on the ^{19}F Spin-Lattice Relaxation Rates of Fluosol-DA, *Phys. Med. Biol.*, **30**, 677-686, 1985.

Rigby BJ, Hirai N, Spikes JD, Eyring H, The Mechanical Properties of Rat Tail Tendon, *J. Gen. Physiol.*, **43**, 265-283, 1958.

Rubin P, Casarett GW, Clinical Radiation Pathology, Vol. 1 and 2, Saunders, Philadelphia, 1968.

Rubin P, Casarett GW, Microcirculation of Tumors, II: The Supervascularized State of Irradiated Regressing Tumors, *Clin. Radiol.*, **17**, 346-355, 1966.

Schwarz SE, Electromagnetics for Engineers, Saunders College Publishing, 1990.

Siemann DW, Hull RP, Bush RS, The Importance of Pre-Irradiation Breathing Times of Oxygen and Carbogen (5% CO_2: 95% O_2) on the *In vivo* Radiation Response of a Murine Sarcoma, *Int. J. Radiat. Oncol. Biol. Phys.*, **2**, 903-911, 1977.

Silva MD, Han SS, Sotak CH, Effects of Signal-to-Noise and Parametric Limitations on Fitting Biexponential Magnetic Resonance (MR) Inversion-Recovery Curves Using a Constrained Nonlinear Least Squares Algorithm, *In. Proc. IEEE NE Bioeng. Conf.*, 24th Annual Meeting, 35-37, Hershey, PA, 1998.

Skach WR, Shi L-B, Calayag MC, Frigeri A, Lingappa VR, Verkman AS, Biogenesis and Transmembrane Topology of the CHIP28 Water Channel at the Endoplasmic Reticulum, *J. Cell. Biol.*, **125**(4), 803-815, 1994.

Smith KL, Daniels JL, Arnoczky SP, Dodds JA, Cooper TG, Gottschalk A, Shaw DA, Effect of Joint Position and Ligament Tension on the MR Signal Intensity of the Cruciate Ligaments of the Knee, *J. Magn. Reson. Imaging*, **4**, 819-822, 1994.

Sostman HD, Rockwell S, Sylvia AL, Madwed D, Cofer G, Charles HC, Negro-Vilar R, Moore D, Evaluation of BA1112 Rhabdomyosarcoma Oxygenation with Microelectrodes, Optical Spectrophotometry, Radiosensitivity, and Magnetic Resonance Spectroscopy, *Magn. Reson. Med.*, **20**, 253-267, 1991.

Sotak CH, Dardzinski BJ, Hees PS, Kaufman RJ, Study of the Biodistribution and Dose Response of Perfluoro-15-Crown-5-Ether in Tumor-Bearing C_3H Mice Using ^{19}F Magnetic Resonance Imaging, *In. Proc. SMRM, 12th Annual Meeting*, p. 983, New York, New York, August 1993.

Sotak CH, Hees PS, Huang HN, Hung MH, Krespan CG, Raynolds S, A New Perfluorocarbon for Use in Fluorine-19 Magnetic Resonance Imaging and Spectroscopy, *Magn. Reson. Med.*, **29**, 188-195, 1993.

Sotak CH, Hees PS, Huang HN, Hung MH, Krespan CG, Raynolds S, Use of PTBD for Murine Tumor Detection with ^{19}F MR Imaging on a 1.5 T Clinical Instrument, *J. Magn. Reson. Imaging*, **1**, 163, 1991.

Stehling MK, Ordidge RJ, Coxon R, Mansfield P, Inversion-Recovery Echo-Planar Imaging (IR-EPI) at 0.5T, *Magn. Reson. Med.*, **13**, 514-517, 1990.

Stejskal EO, Tanner JE, Spin Diffusion Measurements: Spin Echoes in the Presence of a Time-Dependent Field Gradient, *J. Chem. Phys.*, **42**, 288-292, 1965.

Suit HD, Marshall N, Woerner D, Oxygen, Oxygen Plus Carbon Dioxide and Radiation Therapy of a Mouse Mammary Carcinoma, *Cancer*, **30**, 1154-1158, 1972.

Szafer A, Zhong J, Anderson AW, Gore JC, Diffusion Weighted Imaging in Tissues: Theoretical Models, *NMR Biomed.*, **8**, 289-296, 1995.

Tanner JE, Intracellular Diffusion of Water, *Arch. Biochem. Biophys.*, **224(2)**, 416-428, 1983.

Tanner JE, Use of the Stimulated Echo in NMR Diffusion Studies, *J. Chem. Phys.*, **52**, 2523-2526, 1970.

Terris DJ, Minchinton AI, Dunphy EP, Brown JM, Computerized Histographic Oxygen Tension Measurements of Murine Tumors, *Adv. Exp. Med. Biol.*, **317**, 153-159, 1992.

Thomas SR, Clark LC, Ackerman JL, Pratt RG, Hoffman RE, Busse LJ, Kinsey RA, Samaratunga RC, MR Imaging of Lung Using Liquid Perfluorocarbons, *J. Comput. Assist. Tomogr.*, **10**, 1-9, 1986.

Thomlinson RH, Gray LH, The Histological Structure of Some Human Lung Cancers and the Possible Implications for Radiotherapy, *Br. J. Cancer*, **9**, 539-549, 1955.

Trouard TP, Aiken NR, McGovern KA, Correlation of Extracellular Volume Fraction and Apparent Diffusion Coefficient in Red Blood Cell Suspensions *via* DWI and ^{31}P MRS, *In Proc. ISMRM*, p. 1331, 1997.

Trudeau JD, Dixon WT, Hawkins J, The Effect of Inhomogeneous Sample Susceptibility on Measured Diffusion Anisotropy Using NMR Imaging, *J. Magn. Reson. B,* **108**, 22-30, 1995.

Turner R, Le Bihan D, Single Shot Diffusion Imaging at 2.0 Tesla, *J. Magn. Reson.*, **86**, 445-452, 1990.

Twentyman PR, Brown JM, Gray JW, Franko AJ, Scoles MA, Kallman RF, A New Mouse Tumor Model System (RIF-1) for Comparison of Endpoint Studies, *J. Natl. Cancer Inst.*, **64**, 595-604, 1980.

van Dusschoten D, Moonen CTW, de Jager PA, Van As H, Unraveling Diffusion Constants in Biological Tissue by Combining Carr-Purcell-Meiboom-Gill Imaging and Pulsed Field Gradient NMR, *Magn. Reson. Med.*, **36**, 907-913, 1996.

Vaupel P, Hypoxia in Neoplastic Tissue, *Microvasc. Res.*, **13**, 399-408, 1977.

Vétek G, Pályka I, Sotak CH, Springer CS Jr., CR-Free Discrimination of Intra- and Extracellular ^1H_2O Signals from Yeast Cell Suspensions by Diffusion-Space Relaxography (Diffusigraphy), *In. Proc. ISMRM, 2nd Annual Meeting*, p. 1051, San Francisco, CA, 1994.

Viidik A, Lewin T, Changes in Tensile Strength and Histology of Rabbit Ligaments Induced by Different Modes of Postmortal Storage, *Acta Orthop. Scand.*, **37**, 141-155, 1966.

Weast RC, Handbook of Chemistry and Physics, The Chemical Rubber Co., Cleveland, Ohio, 1971.

Curriculum Vitae

Sam S. Han, Ph.D., J.D.

2095 Raceway Trail
Beavercreek, OH 45434
(937) 401-0070 • sam.han.esq@gmail.com

EDUCATION

Georgia State University   *Atlanta, GA*
- **J.D. (cum laude)**   May, 2001
  - *Class Rank 11 of 163 (Average = 86.50)*
  - *Moot Court*

Worcester Polytechnic Institute   *Worcester, MA*
- **M.E. (Biomedical Engineering)**   February, 1998
- **Ph.D. (Biomedical Engineering)**   May, 1998

The University of Michigan   *Ann Arbor, MI*
- **B.S.E. (Electrical Engineering)**   August, 1992
  Worked full time concurrent with full time academics

ADMISSIONS

- Federal Courts of Appeal
  - The Supreme Court of the United States
  - The Court of Appeals for the Sixth Circuit
  - The Court of Appeals for the Eleventh Circuit
  - The Court of Appeals for the Federal Circuit

- Admitted to practice law in the State of Georgia
  - Georgia Bar Number 322284
    - The Court of Appeals for the State of Georgia
    - The Supreme Court of Georgia
    - The United States District Court for the Northern District of Georgia

- Admitted to practice before the United States Patent and Trademark Office
  - Registration Number 51,771

WORK EXPERIENCE

Of Counsel   McClure, Qualey & Rodack, LLP, Atlanta, GA
(2011 - Present)
- Patent
  - Applications
  - Responses to Office Action
  - Appeals
  - Opinions of Counsel

Curriculum Vitae
Sam S. Han, Ph.D., J.D.

2095 Raceway Trail
Beavercreek, OH 45434
(937) 401-0070 • sam.han.esq@gmail.com

- Trademark
  - Applications
  - Opposition Proceedings
  - Appeals

Consultant   *DeWoskin Law Firm, LLC, Atlanta, GA*
(2008 - Present)
- Advise on Fair Debt Collection Practices Act
- Advise on Georgia Practices and Procedures for Superior Court and State Court
- Advise on Federal Rules of Civil Procedure and Local Civil Rules for the Northern District of Georgia

Of Counsel   *Thomas, Kayden, Horstemeyer, & Risley, LLP, Atlanta, GA*
(2010 - 2011)
- Patent
  - Applications
  - Responses to Office Action
  - Appeals

Adjunct Professor of Law   *Indiana Institute of Technology, Fort Wayne, IN*
(2015)
- Publications (*infra*).
- Teaching and Instruction (*infra*).
- Presentations and Conference Proceedings (*infra*).
- Lectures and Speaking Engagements (*infra*).
- Books and Chapters (*infra*).
- Academic Advising (*infra*).
- Fellowships, Scholarships, and Grants (*infra*).
- Awards and Recognitions (*infra*).
- Academic Committees and Service (*infra*).

Assistant Professor of Law   *The University of Dayton, School of Law, Dayton, OH*
(2008 - 2012)
- Publications (*infra*).
- Teaching and Instruction (*infra*).
- Presentations and Conference Proceedings (*infra*).
- Lectures and Speaking Engagements (*infra*).
- Books and Chapters (*infra*).
- Academic Advising (*infra*).
- Fellowships, Scholarships, and Grants (*infra*).
- Awards and Recognitions (*infra*).
- Academic Committees and Service (*infra*).

Curriculum Vitae

Sam S. Han, Ph.D., J.D.

2095 Raceway Trail
Beavercreek, OH 45434
(937) 401-0070 • sam.han.esq@gmail.com

Chief Litigation Counsel Furukawa Electric North America, Inc. ("FENA"), Norcross, GA
(2007 - 2008)
- Example litigation matters
    - *Furukawa Electric North America, Inc. v. Alcatel, Alcatel NA Cable Systems, Inc. d/b/a Alcatel Telecommunications Cable, and Alcatel USA*, C.A. No. 5:03-CV-111-V (W.D.N.C.)
    - *In re FiberCore*, Chapter 11, Case No. 03-46551-HJB, (U.S. Bankr. W.D. Mass.)
    - *Steven Weiss, Trustee on Behalf of FiberCore, Inc. v. OFS Fitel, LLC and Furukawa Electric North America, Inc., f/k/a Fitel USA Corp.*, Adversary Action No. 06-04168-HJB, (U.S. Bankr. W.D. Mass.)
    - *Fitel USA Corp. v. FiberCore, Inc.*, C.A. No. 5:02-CV-163-V (W.D.N.C.)
    - *Furukawa Electric North America, Inc. v. Sterlite Optical Technologies, Ltd., Sterlite Optical Technologies, Inc., Anand Agarwal, and Brian Chomniak*, Civil Action No. 1:02-CV-2149-CAP (N.D. Ga.)
    - *Furukawa Electric North America, Inc. and OFS Fitel, LLC v. Nufern, Inc.*, Civil Action Number 03:05-CV-01252 (MRK) (D. Conn.)
    - *Furukawa Electric North America, Inc. and OFS Fitel, LLC v. Yangtze Optical Fibre and Cable Company, Ltd.*, Civil Action Number 05-11219 (RGS) (D. Mass.)
    - *Furukawa Electric North America, Inc. v. Draka Comteq Americas, Inc., et al.*, Civil Action Number 05:07-CV-00058 (RLV) (W.D.N.C.)
- Legal Counsel
    - Nondisclosure and Confidentiality Agreement for Furukawa Electric North America, Inc.
    - Licensing negotiations and licensing agreements relating to the intellectual property portfolio and intangible assets of Furukawa Electric North America, Inc.
    - Validity/invalidity analyses for patents in portfolio of Furukawa Electric North America, Inc.
    - Freedom-to-use and clearance analyses for business units owned by Furukawa Electric North America, Inc.
    - Negotiation and execution of purchase orders for software upgrades
    - Negotiation and execution of joint development agreements
    - Advise on taxable consequences of transfer of intangible assets
    - Manage trademark portfolio for Furukawa Electric North America, Inc., and OFS Fitel, LLC

Intellectual Property Counsel OFS Fitel, LLC (Subsidiary of FENA), Norcross, GA
(2006)
- Example intellectual property litigation matters
    - *Furukawa Electric Company of North America, Inc. and OFS Fitel, LLC v. Nufern, Inc.*, Civil Action Number 03:05-CV-01252 (MRK) (D. Conn.)
    - *Furukawa Electric Company of North America, Inc. and OFS Fitel, LLC v. Yangtze Optical Fibre and Cable Company, Ltd.*, Civil Action Number 05-11219

Curriculum Vitae
Sam S. Han, Ph.D., J.D.
2095 Raceway Trail
Beavercreek, OH 45434
(937) 401-0070 • sam.han.esq@gmail.com

(RGS) (D. Mass.)
- Legal Counsel
    - Nondisclosure and Confidentiality Agreements for OFS Fitel, LLC, and various entities
    - Licensing negotiations and licensing agreements
    - Infringement/non-infringement analyses for technology developed by OFS Fitel, LLC
    - Validity/invalidity analyses for technology developed by OFS Fitel, LLC

Associate (Commercial Litigation)　　　　　　　　　　　*McGuireWoods LLP, Atlanta, GA*
(2005 - 2006)
- Example Commercial Litigation Matters
    - *Thomas v. Alltel Communications, Inc., et al.*, No. 3:05cv506 (W.D.N.C.)
    - *Digital Envoy, Inc. v. Google, Inc.*, No. C-04-01497-RS (N.D. Cal.)
    - *CBeyond Communications, LLC v. Reignmaker Consortium et al.*, No. 05-1-03939-18 (Superior Court, Cobb Co., Georgia)
    - *Gardendance, Inc. v. Woodstock Copperworks, Ltd.*, No. 1:04-CV-00010 (M.D.N.C.)
    - *Moses v. Traton Corp.*, No. 05-1-08395-35 (Superior Court, Cobb Co., Georgia)
- Example Pro Bono and Public Service Matters
    - Legal consultation role in *Pressley v. Metropolitan Atlanta Rapid Transit Authority, et al.*, 2003CV76160 (Superior Court, Fulton Co., Georgia)
    - Legal consultation role in *Duong et al. v. Hyun et al.*, Case No. SCN-107635 (Small Claims Court, San Mateo Co., California)
    - Legal consultation role in *Hyun v. Duong, et al.*, Case No. CIV 449385 (Superior Court, San Mateo Co., California)
- Legal Counsel
    - Nondisclosure and Confidentiality Agreement between Georgia corporation and Illinois corporation
    - Review and advise on open source licensing and ramifications on proprietary software
    - Research and advising on newly-enacted laws related to wholesalers and manufacturers in the pharmaceutical industry
    - Research and analyses on trade dress matters
    - Mediation for civil litigation
    - Commercial arbitration matters
    - Witness interviews for class action lawsuit
    - Breach of contract litigation matters
- Intellectual Property-Related Matters
    - Patent application for invention related to vinyl siding
    - Patent applications for invention related to waste reclamation and recycling
    - Provisional patent application for invention related to double-knuckle door hinges
    - *Markman* Brief for lighting fixtures-related litigation matter

Curriculum Vitae
Sam S. Han, Ph.D., J.D.

2095 Raceway Trail
Beavercreek, OH 45434
(937) 401-0070 • sam.han.esq@gmail.com

- Deposition of witness in relation to claim construction
- Deposition of witness in relation to infringement issues
- Review and analyses of patents related to gussett rufflers for manufacturing beds
- Abbreviated New Drug Application matter
- Provocation of Interference with the United States Patent and Trademark Office (USPTO)
- Trademark prosecution matters

Associate   Thomas, Kayden, Horstemeyer, & Risley, LLP, Atlanta, GA
(2001 - 2005)
- Highest billing associate (2003)
- Liaison to various South Korean clients
- Patent
  - Applications
  - Responses to Office Action
  - Patentability Opinions
  - Infringement/Noninfringement Analyses and Opinions
  - Validity/Invalidity Analyses and Opinions
  - Enforceability Analyses
  - Patent Clearance Searches
- Trademark
  - Applications
  - Office Action Responses
- Licensing projects
  - Analysis and licensing efforts for patent portfolio of Datascape, Inc.
  - Analysis and licensing efforts for patent portfolio of Paradyne Corp.
  - Analysis and licensing efforts for patent portfolio of Scientific Atlanta
- Intellectual property-related litigation matters
  - *Datascape v. FleetBoston Financial Corp.*, No.: 1:01-CV-2246-CAP (N.D. Ga.)
  - *Datascape v. Providian Financial Corp.*, No.: 1:01-CV-3252-CAP (N.D. Ga.)
  - *Datascape v. Visa USA, Inc.*, No.: 1:02-CV-0289-CAP (N.D. Ga.)
  - *Safer Display Tech., Ltd. v. Proview Electr. Co., Ltd.*, Case No. 2:04cv161 (E.D. Va.)
- Pro bono and public service matters
  - Legal consultant on *State v. Pressley*, A04A1197 (Court of Appeals, Georgia)
  - Legal consultant on *State v. Pressley*, 02-CR-271323 (State Court, Fulton Co., Georgia)
  - Non-legal consultation role in *Pressley v. Metropolitan Atlanta Rapid Transit Authority, et al.*, 2003CV76160 (Superior Court, Fulton Co., Georgia)

Technical Advisor   Thomas, Kayden, Horstemeyer, & Risley, LLP, Atlanta, GA
(1999 - 2001)
- Patent

Curriculum Vitae
Sam S. Han, Ph.D., J.D.
2095 Raceway Trail
Beavercreek, OH 45434
(937) 401-0070 • sam.han.esq@gmail.com

- Applications
- Responses to Office Action
- Patentability Opinions
  - Trademark
    - Response to Office Action
    - Infringement Opinion Letter
    - Cease and Desist Letter
  - Intellectual property-related litigation research
    - Issues on Willful Patent Infringement
    - Issues Regarding the Waiver of Attorney Client Privilege and Infringement Opinion
    - Trademark Infringement
    - Bankruptcy Issues Related to Intellectual Property as Collateral
    - Contract and Tort Damages in Intellectual Property Litigation

Extern *Hon. Marvin Shoob, U. S. District Court, Northern District of Georgia, Atlanta, GA* (2000 Fall Term)
- Legal Research
  - Employment Discrimination
  - First Amendment
  - Takings and the Fourteenth Amendment
  - Corporate Law
- Observation of Courtroom Proceedings
  - Plea Hearings
  - Sentencing Hearings
  - Probation Revocation Hearings
  - Daubert Hearing
- Writing Orders for the Court
  - Motion for Summary Judgment in Personal Injury Diversity Case
  - Motion for Summary Judgment in Employment Discrimination Case
  - Motion for Summary Judgment in First Amendment Case
  - Motion for Summary Judgment in Corporate Fraud Case

Summer Associate *Thomas, Kayden, Horstemeyer, & Risley, LLP, Atlanta, GA* (2000 Summer)
- Patent
  - Applications
  - Preliminary Amendment
  - Response to Office Action
  - Request for Correction
- Trademark
  - Response to Office Action
  - Opinion Letter

- Cease and Desist Letter
- Intellectual property related litigation research
  - Issues on Willful Patent Infringement
  - Issues Regarding the Waiver of Attorney Client Privilege and Infringement Opinion
  - Trademark Infringement
  - Bankruptcy Issues
  - Contract and Tort Damages in Intellectual Property Litigation

Extern   Hon. Wendy L. Shoob, Fulton County Superior Court, Atlanta, GA
(2000 Spring Term)
- Legal Research
  - Divorce
  - Corporate Contract
  - Bankruptcy
  - Appeals from Administrative Proceedings
  - Evidentiary Issues in Criminal Proceedings
- Observation of Courtroom Proceedings
  - Plea and Arraignment
  - Voir Dire of Jury
  - Pre-Trial Conference
  - Criminal and Civil Proceedings
  - Writing Orders for the Court
- Writing Orders for the Court
  - Corporation Contracts
  - Appeal from Administrative Proceedings

Summer Associate   Jones & Askew, LLP, Atlanta, GA
(1999 Summer)
- Patent
  - Disclosure Meeting with Client
    - Application
    - Response to Office Action
    - Information Disclosure Statement
    - Opinion Letter
- Trademark
  - Response to Office Action
  - Amendment to the Supplemental Register
  - Opinion Letter
- Intellectual property related litigation research
  - Patent Infringement
  - Trademark Infringement
  - Internet Domain Name and Trademark Dispute

Curriculum Vitae
Sam S. Han, Ph.D., J.D.
2095 Raceway Trail
Beavercreek, OH 45434
(937) 401-0070 • sam.han.esq@gmail.com

▫ Trade Secret

Research Scientist / Research Assistant Worcester Polytechnic Institute, Worcester, MA
(1993 - 1998)
- Scientific presentations and publications
- Design and execution of original scientific and medical experiments
- Collaborative efforts in medicine and experimental biomedicine
- Responsible for animal care and handling

Teaching Assistant / Instructor Worcester Polytechnic Institute, Worcester, MA
(1994 - 1998)
- Responsible for lab classes and lectures in the Biomedical Engineering Department
- Teaching responsibilities for graduate and undergraduate classes

Systems Administrator WPI NMR Research Group, Worcester, MA
(1994 - 1998)
- Responsible for maintaining UNIX workstations
- Assisted in writing custom C and IDL code for image processing and analysis
- Integration of HP, SUN, and Silicon Graphics Workstations with PC-based networks

Quality Assurance Engineer Central Massachusetts Magnetic Imaging Center, Worcester, MA
(1993 - 1997)
- Responsible for maintaining high level of quality on clinical MRI scanners
- Liaison between clinical branch and research branch of Central Massachusetts Magnetic Imaging Center

Counselor / Teacher Interlochen Center for the Performing Arts, Traverse City, MI
(1991 and 1992 Summer)
- Responsible for the health and well being of students
- Teaching responsibilities

Lab Assistant University of Michigan Department of Epidemiology, Ann Arbor, MI
(1988 - 1990)
- Lab equipment maintenance
- Cell cultures and hazardous materials

LITIGATION AND TRIAL EXPERIENCE

- *Abbott Laboratories, Inc. v. Mylan Pharmaceuticals, Inc.*, No. 05 C 6561 (N.D. Ill.)
- *Abbott Laboratories, Inc. v. Mylan Pharmaceuticals, Inc.*, No. 1:05-CV-150 (N.D.W.V.)
- *Arrow Financial Services, LLC v. Wright*, No. 08-VS-130387-F (State Court, Fulton Co., Georgia)

Curriculum Vitae
Sam S. Han, Ph.D., J.D.

2095 Raceway Trail
Beavercreek, OH 45434
(937) 401-0070 • sam.han.esq@gmail.com

- *CACH, LLC v. Divergilio*, No. 08-VS-136117-C (State Court, Fulton Co., Georgia)
- *CACH, LLC v. Cain*, No. 08-A-90282-3 (State Court, DeKalb Co., Georgia)
- *Carmel Holdings I, LLC v. Sandlin*, No. 08-VS-134346 (State Court, Fulton Co., Georgia)
- *CBeyond Communications, LLC v. Reignmaker Consortium et al.*, No. 05-1-03939-18 (Superior Court, Cobb Co., Georgia)
- *Chattman v. Alpha Receivables, Inc., et al.*, 1:09-CV-1525-GET-ECS (N.D. Ga.)
- *Coupons, Inc. v. Value America, Inc.*, No. 1:99-CV-0758 (N.D. Ga.)
- *Datascape v. FleetBoston Financial Corp.*, No. 1:01-CV-2246-CAP (N.D. Ga.)
- *Datascape v. Providian Financial Corp.*, No. 1:01-CV-3252-CAP (N.D. Ga.)
- *Datascape v. Visa USA, Inc.*, No. 1:02-CV-0289-CAP (N.D. Ga.)
- *Dieujuste v. Onyx Waste Services Southeast, Inc.*, Case No. 04-81104 CIV (S.D. Fl., West Palm Beach Division)
- *Digital Envoy, Inc. v. Google, Inc.*, No. C-04-01497-RS (N.D. Cal.)
- *Dippin' Dots, Inc. v. Concession Obsession, Inc.*, No. Civil 1:00-Cv-907-TWT (N.D. Ga.)
- *Draka-Comteq Americas, Inc. v. Furukawa Electric North America, Inc. and OFS Fitel, LLC,*, No. 2:07-CV-00352-LED (E.D. Tex.)
- *Draka-Comteq Americas, Inc. v. Furukawa Electric North America, Inc. and OFS Fitel, LLC,*, No. 2:07-CV-00427 (E.D. Tex.)
- *ESHARE Technologies, Inc., v. Stratasoft, Inc.*, No. 1:99-CV-2303-RWS (N.D. Ga.)
- *European American Realty, Ltd. and Scott K. Toberman v. David Lang*, No. 2005-CV-105849 (Superior Court, Fulton Co., Georgia)
- *Evans v. Givins*, 06-M-36010 (Magistrate Court, Gwinnett Co., Georgia)
- *Fine & Assoc., P.C. v. Jones*, No. 09-M-16283 (Magistrate Court, Gwinnett Co., Georgia)
- *Fitel USA Corp. v. FiberCore, Inc.*, C.A. No. 5:02-CV-163-V (W.D.N.C.)
- *Frampton & Assoc., Inc. v. Ceco Building Systems*, Case No. 2004-CP-10-1877 (Ct. Common Pleas S.C., 9th Jud. Cir., Charleston Co., South Carolina)
- *Furukawa Electric North America, Inc. and OFS Fitel, LLC v. Nufern, Inc.*, Civil Action Number 03:05-CV-01252 (MRK) (D. Conn.)
- *Furukawa Electric North America, Inc. and OFS Fitel, LLC v. Yangtze Optical Fibre and Cable Company, Ltd.*, Civil Action Number 05-11219 (RGS) (D. Mass.)
- *Furukawa Electric North America, Inc. v. Alcatel, Alcatel NA Cable Systems, Inc. d/b/a Alcatel Telecommunications Cable, and Alcatel USA*, C.A. No. 5:03-CV-111-V (W.D.N.C.)
- *Furukawa Electric North America, Inc. v. Draka Comteq Americas, Inc., et al.*, Civil Action Number 05:07-CV-00058 (RLV) (W.D.N.C.)
- *Furukawa Electric North America, Inc. v. Sterlite Optical Technologies, Ltd., Sterlite Optical Technologies, Inc., Anand Agarwal, and Brian Chomniak*, Civil Action No. 1:02-CV-2149-CAP (N.D. Ga.)
- *Gardendance, Inc. v. Woodstock Copperworks, Ltd.*, No. 1:04-CV-00010 (M.D.N.C.)
- *Han v. The University of Dayton, et al.*, Civil Action No. 2011-CV-08966 (Ct. Common Pleas, Montgomery Co., Ohio)
- *Humbert v. The City of College Park, et al.*, Civil Action No. 1:05-cv-2740-GET (N.D. Ga.)
- *In re FiberCore, Inc.*, Chapter 11, Case No. 03-46551-HJB, (U.S. Bankr. W.D. Mass.)

Curriculum Vitae
Sam S. Han, Ph.D., J.D.
2095 Raceway Trail
Beavercreek, OH 45434
(937) 401-0070 • sam.han.esq@gmail.com

- *Johnston Industries Composite Reinforcements, Inc. v. V2 Composite Reinforcements, Inc.*, No. 00-T-728-E (M.D. Ala.)
- *Kamitani v. Lite-On, Inc.*, No. H-01-4123 (S.D. Tex.)
- *Lavender v. Lavender*, No. 07-CV-2211-B (Superior Court, Barrow Co., Georgia)
- *LVNV Financial, LLC v. Gordon*, No. 08-VS-137017 (State Court, Fulton Co., Georgia)
- *Moses v. Traton Corp. et al.*, No. 05-1-08395-35 (Superior Court, Cobb Co., Georgia)
- *Moses v. Traton Corp. et al.*, No. 06-1-08441-35 (Superior Court, Cobb Co., Georgia)
- *NCO Portfolio Management, Inc. v. Jessica N. Buonodono*, No. 2007-A-2264-5 (State Court, Cobb Co., Georgia)
- *Noguchi v. Cooper, et al.*, No. 07-MS-076113 (Magistrate Court, Gwinnett Co., Georgia)
- *Portfolio Recovery Associates, LLC v. Loncar*, No. 08-VS-130682 (State Court, Fulton Co., Georgia)
- *Pressley v. Metropolitan Atlanta Rapid Transit Authority, et al.*, No. 2003-CV-76160 (Superior Court, Fulton Co., Georgia)
- *Prochroma Technologies, Inc. v. U.S.*, No. 97-139C (Ct. Cl.)
- *Pyen v. Kimsey, et al.*, No. 08A00899-8 (Superior Court, Gwinnett Co., Georgia)
- *Raines v. Roach*, No. 2006-CV-126099 (Superior Court, Fulton Co., Georgia)
- *Robinson v. Correctional Medical Assoc., et al.*, No. 2009-CV-167587 (Superior Court, Fulton Co., Georgia)
- *Robinson v. Correctional Medical Assoc., et al.*, No. 1:09-cv-01509 (N.D. Ga.)
- *Safer Display Tech., Ltd. v. AmTRAN Technology Corp.*, No. 2:04cv154 (E.D. Va.)
- *Safer Display Tech., Ltd. v. Proview Electr. Co., Ltd.*, No. 2:04cv161 (E.D. Va.)
- *State v. Han*, 06-T-9510 (State Court, Cobb Co., Georgia)
- *State v. Jeon*, 07-T-16826 (State Court, Cobb Co., Georgia)
- *State v. Pressley*, 02-CR-271323 (State Court, Fulton Co., Georgia)
- *Steven Weiss, Trustee on Behalf of FiberCore, Inc. v. OFS Fitel, LLC and Furukawa Electric North America, Inc., f/k/a Fitel USA Corp.*, Adversary Action No. 06-04168-HJB, (U.S. Bankr. W.D. Mass.)
- *Stringer v. Gordon*, No. 09-ED-426587 (Magistrate Court, Fulton Co., Georgia)
- *Thomas v. Alltel Communictions, Inc., et al.*, No. 3:05cv506 (W.D.N.C.)
- *Toshiba Corporation v. Ultima Electronics Corp.*, No. CV-02-04825 CAS (C.D. Cal.)
- *Traton News, LLC, v. Traton Corp., et al.*, No. 3:11cv00435-WHR (S.D. Ohio)
- *Unifund Partners v. Stinyard*, No. 08-VS-135327-G (State Court, Fulton Co., Georgia)

APPELLATE EXPERIENCE
- *State v. Pressley*, A04A1197 (Court of Appeals, Georgia)
- *Moses v. Traton Corp. et al.*, S07A0780 (Supreme Court, Georgia)
- *Moses v. Traton Corp. et al.*, S07C1858 (Supreme Court, Georgia)
- *Moses v. Traton Corp. et al.*, A07A1474 (Court of Appeals, Georgia)

LEGAL COUNSELING, LICENSING, AND OPINIONS

Curriculum Vitae

Sam S. Han, Ph.D., J.D.

2095 Raceway Trail
Beavercreek, OH 45434
(937) 401-0070 • sam.han.esq@gmail.com

- Standards Coverage Analysis
  - ITU G.DMT (Discrete Multi-Tone) Standard (ITU-G.992.1): Asymmetrical Digital Subscriber Line (ADSL) Transceivers
  - T1.413 Issue 2 Standard: Asynchronous Transfer Mode (ATM) Based Asymmetric Digital Subscriber Line (ADSL) Internet Connection Sharing (ICS)
  - DOCSIS 1.0 Standard: Data-Over-Cable Service Interface Specification
  - DOCSIS 1.1 Standard: Data-Over-Cable Service Interface Specification
  - DOCSIS 2.0 Standard: Data-Over-Cable Service Interface Specification
  - ITU-T G.729 Standard: Coding of Speech at 8 kbits/s Using Conjugate-Structure Algebraic-Code-Excited Linear-Prediction (CS-ACELP)
- Agreement on Division of Intellectual Property for Joint Research
- Intellectual Property Counsel for Various General Practice Attorneys
- Agreements on Joint Development, Non-Disclosure, and Intellectual Property Division
  - Furukawa Electric North America, Inc.
  - OFS Laboratories, LLC
  - OFS Fitel, LLC
- Licensing Negotiations
- Patentability Searches and Opinions
- Infringement/Noninfringement Opinions
- Validity/Invalidity Searches and Opinions
- Freedom-to-Operate (Clearance) Searches and Opinions
- Analysis of Contested Cases
- Inventorship Determinations
- Trademark Opposition Proceedings
- Trademark Cancellation Proceedings
- Domain Names and Internet-Related Issues

PATENT AND TRADEMARK PROSECUTION MATTERS

- Australian Patent Office
- Chinese Patent Office
- European Patent Office (EPO)
- French Patent Office
- Japanese Patent Office (JPO)
- Korean Intellectual Property Office (KIPO)
- Patent Applications Filed Under the Patent Cooperation Treaty (PCT)
- Taiwanese Patent Office
- United States Patent and Trademark Office (USPTO) Proceedings
  - Appeals Before the Board of Patent Appeals and Interferences
  - Reexamination Requests and Proceedings
  - Issued Design Patents

Curriculum Vitae

Sam S. Han, Ph.D., J.D.

2095 Raceway Trail
Beavercreek, OH 45434
(937) 401-0070 • sam.han.esq@gmail.com

- Design Patent Applications
- Non-Provisional Utility Patent Applications
- Provisional Patent Applications
- Trademark Applications and Continued Prosecution
- Continued Prosecution of Trademark Portfolio of Furukawa Electric North America, Inc.
- Trademark Opposition Proceedings

PUBLICATIONS

- S. S. Han, Association of Molecular Pathology *Meets* Therasense: *Analyzing the Unenforceability of Isolated-Sequence-Related Patents for UPenn, Columbia, NYU, Yale, and Emory*, 17 J. Tech. L. & Pol'y 1 (2012).
- S. S. Han, Therasense *Nonsense*, 37 U. Dayton L. Rev. 185 (2012).
- S. Ganow, S. S. Han, *Model Omnibus Privacy Statute*, 35 U. Dayton L. Rev. 345 (2010).
- S. S. Han, *Daubert v. Markman: Fact Experts on Issues that are Wholly Devoid of Any Factual Component*, 50 IDEA: J. L. & Tech 367 (2010).
- S. S. Han, *Predicting the Enforceability of Browse-Wrap Agreements in Ohio*, 36 Ohio N. U. L. Rev. 1 (2010).
- S. S. Han, *Analyzing the Patentability of "Intangible" Yet "Physical" Subject Matter*, 3 Colum. Sci. & Tech. L. Rev. 2 (February 19, 2001) <http://www.stlr.org/cite.cgi?volume=3&article=2>.
- J. J. Neil, T. Q. Duong, C. S. Springer, C. H. Sotak, G. L. Bretthorst, I. Pályka, G. Vétek, S. S. Han, J. J. H. Ackerman, *Evaluation of Equilibrium Transcytolemmal Water Exchange in Rat Brain*, Submitted, Science.
- M. D. Silva, K. G. Helmer, J-H Lee, S. S. Han, C. S. Springer, C. H. Sotak, *Deconvolution of Compartmental Water Apparent Diffusion Coefficient in Yeast-Cell Suspensions Using Combined T_1 and Diffusion Measurements*, J. Magn. Reson. **156**, 52-63 (2002).
- Han, S., Gemmell S. J., Helmer, K. G., Grigg, P., Wellen, J. W., Hoffman, A. H., Sotak, C. H., *Changes in ADC Caused by Tensile Loading of Rabbit Achilles Tendon: Evidence for Water Transport*, J. Magn. Reson. **144(2)**, 217 - 227 (2000).
- F. Li, S. S. Han, T. Tatlisumak, K-F. Liu, J. H. Garcia, C. H. Sotak, M. Fisher, *Reversal of Acute Average Apparent Diffusion Coefficient Abnormalities and Delayed Neuronal Damage Following Varying Periods of Transient Focal Cerebral Ischemia in Rats*, Annals of Neurology **46(3)**, 333-342 (1999).
- F. Li, S. S. Han, T. Tatlisumak, R. A. D. Carano, K. Irie, C. H. Sotak, M. Fisher, *A New Method to Improve the In-Bore Middle Cerebral Artery Occlusion: Demonstration with Diffusion- and Perfusion-Weighted Imaging*, Stroke **29**, 1715 - 1720 (1998).
- K. G. Helmer, S. S. Han, C. H. Sotak, *Correlation Between the Apparent Diffusion Coefficient of Tissue Water and Oxygen Tension in RIF-1 Tumors*, NMR in Biomedicine **11**, 120 - 130 (1998).

TEACHING AND INSTRUCTION

- Indiana Institute of Technology, School of Law, Fort Wayne, IN

Curriculum Vitae

Sam S. Han, Ph.D., J.D.

2095 Raceway Trail
Beavercreek, OH 45434
(937) 401-0070 • sam.han.esq@gmail.com

- 2015 Fall
    - *Intellectual Property.*

- The University of Dayton, School of Law, Dayton, OH
    - 2012 Spring
        - *Patent Litigation* (LAW 6905-01).
        - *Licensing Intellectual Property* (LAW 6420-01).
        - *Moot Court* (LAW 6872-01).
        - *Civil Engineering Seminar* (CEE 300), April 28, 2011.
        - *Project Management and Innovation* (MEE 433 / ECE 433), March 17, 2011.
    - 2011 Fall
        - *Patent Law* (LAW 6425-01).
        - *Patent Practice and Procedure* (LAW 6940-01).
        - *Moot Court* (LAW 6872-01)
        - *Civil Engineering Seminar* (CEE 300), November 3, 2011.
        - *Project Management and Innovation* (MEE 433 / ECE 433), November 1, 2011.
    - 2011 Spring
        - *Intellectual Property* (LAW 6400-01).
        - *Licensing Intellectual Property* (LAW 6420-01).
        - *Moot Court* (LAW 6872-01)
        - *Civil Engineering Seminar* (CEE 300), April 28, 2011.
        - *Project Management and Innovation* (MEE 433 / ECE 433), March 17, 2011.
    - 2010 Fall
        - *Patent Law* (LAW 6425-01).
        - *Patent Practice and Procedure* (LAW 6940-01).
        - *Dot.com Law* (LAW 6943-01)
        - *Moot Court* (LAW 6872-01)
        - *Project Management and Innovation* (MEE 433 / ECE 433), September 30, 2010.
    - 2010 Spring
        - *Licensing Intellectual Property* (LAW 6420-01).
        - *Patent Litigation* (LAW 6905-01)
        - *Project Management and Innovation* (MEE 433 / ECE 433), February 18, 2010.
    - 2009 Fall
        - *Patent Law* (LAW 6425-01).
        - *Patent Practice and Procedure* (LAW 6940-01).
        - *Project Management and Innovation* (MEE 433 / ECE 433), October 23, 2009.
    - 2009 Spring
        - *Licensing Intellectual Property* (LAW 6420-01).
        - *Patent Practice and Procedure* (LAW 6940-01).
        - *Intellectual Property Law* (Law 6400), March 10, 2009.
        - *Intellectual Property Law* (Law 6832), February 3, 2009.
    - 2008 Fall
        - *Patent Law* (LAW 6425-01).

Curriculum Vitae
Sam S. Han, Ph.D., J.D.
2095 Raceway Trail
Beavercreek, OH 45434
(937) 401-0070 • sam.han.esq@gmail.com

- - *Project Management and Innovation* (MEE 433 / ECE 433), November 14, 2008.
  - *Intellectual Property Law* (LAW 6400), September 25, 2008.
- Worcester Polytechnic Institute, Department of Biomedical Engineering, Worcester, MA
  - 1994-1998 (Teaching Assistant / Instructor)
    - *Principles of In Vivo Nuclear Magnetic Resonance* (BME 582).
    - *Laboratory Animal Surgery* (BME 562).
    - *Biomedical Instrumentation* (BME 523).
    - *Biomedical Signal Analysis* (BME 4011).

PRESENTATIONS AND CONFERENCE PROCEEDINGS

- C. R. Miller, S. S. Han, *ENCODE: The End of an Established Era*, Dayton Intellectual Property Law Association, Dayton, OH, September 13, 2013.
- S. S. Han, C.R. Miller, *Gene: the New Four Letter Word*, CincyBio 2013, Cincinnati, OH, April 30, 2013.
- S. S. Han, *Substance Abuse*, Continuing Legal Education, The University of Dayton, School of Law, Dayton, OH, May 19, 2012.
- S. S. Han, H. Farrell, *Dynamics for Personal Motivation and Chemical Addiction*, Ohio State Bar Association, Annual Convention, Cincinnati, OH, May 3, 2012.
- S. S. Han, *A Confrontation of Epic Proportions:* Prometheus v. Mayo, CincyBio 2012, Cincinnati, OH, April 17, 2012.
- S. S. Han, *Biochemical Pathways: A Presentation on Substance Abuse*, Toledo Intellectual Property Law Association, The University of Toledo College of Law, Toledo, OH, March 7, 2012.
- S. S. Han and J. Huebert, *Do We Need Copyrights and Patents?*, Federalist Society, The University of Dayton School of Law, Dayton, OH, February 27, 2012.
- S. S. Han, Therasense *Nonsense*, Ohio Legal Scholarship Workshop, Capital University Law School, Columbus, OH, February 25, 2012.
- S. S. Han, Commenter for Presentation by K. A. Behre on *Motivation for Law Student Pro Bono Work: Lessons Learned from the Tuscaloosa Tornado*, Ohio Legal Scholarship Workshop, Capital University Law School, Columbus, OH, February 25, 2012.
- S. S. Han, *Human Decision-Making and Substance Abuse*, Dayton Intellectual Property Law Association, The University of Dayton, Dayton, OH, October 14, 2011.
- S. S. Han, Association of Molecular Pathology *Meets* Therasense*: Analyzing the Unenforceability of Isolated-Sequence-Related Patents for UPenn, Columbia, NYU, Yale, and Emory*, Faculty Colloquium, The University of Dayton, Dayton, OH, October 3, 2011.
- S. S. Han, H. Farrell, *Substance Abuse - Addicted Attorneys Should Understand the Dynamics that Drive Motivation, Success, and Chemical Dependency*, The 21st All Ohio Annual Institute on Intellectual Property, Cleveland and Cincinnati, OH, September 20-21, 2011.
- S. S. Han, *Therasense Meets the ACLU*, Toledo Intellectual Property Law Association, The Toledo Club, Toledo, OH, September 15, 2011.
- S. S. Han, *The Unenforceability of Isolated-Sequence-Related Patents for Yale, Emory, UPenn, Columbia, and NYU*, Intellectual Property Scholars Conference 2011, DePaul University College

Curriculum Vitae
Sam S. Han, Ph.D., J.D.
2095 Raceway Trail
Beavercreek, OH 45434
(937) 401-0070 • sam.han.esq@gmail.com

of Law, Chicago, IL, August 10-12, 2011.
- S. S. Han, *The Unenforceability of Gene-Related Patents for Yale, Emory, UPenn, Columbia, and NYU*, Ohio Legal Scholarship Workshop, Cleveland-Marshall College of Law, Cleveland, OH, June 23, 2011.
- S. S. Han, Commenter for Presentation by P. Lee on *Innovation and Entrepreneurship in Evolving Economies: The Role of Law*, Ohio Legal Scholarship Workshop, Clevaland-Marshall College of Law, Cleveland, OH, June 23, 2011.
- S. S. Han, *Are the Offspring of Harvard Mice the Product of Unprotected Sex?*, Continuing Legal Education, The University of Dayton, School of Law, Dayton, OH, May 14, 2011.
- S. S. Han, Invited Conference Fellow, *Some Modest Proposals 4.0: A Conference About Pouring Academic Ideas into Legislative Bottles*, Cardozo IP & Information Law Program, Benjamin N. Cardozo School of Law, Yeshiva University, NY, April 8, 2011.
- S. S. Han, *Man v. Nature: Patentability of Second-Generation Biological Organisms*, CincyBio 2011, Cincinnati, OH, March 8, 2011.
- S. S. Han, *Nature Abhors a Vacuum: Prophesying Why China Will Fill the Sucking Void Created by the Collapse of North Korea*, Ohio Legal Scholarship Workshop, Capital University Law School, Columbus, OH, February 5, 2011.
- S. S. Han, T. Reilly, J. Zink, M. Lampke, *Scholarly Presentations on Issues Related to Intellectual Property Law*, Dayton Intellectual Property Law Association, The University of Dayton, School of Law, Dayton, OH, January 13, 2011.
- S. S. Han, *Are the Offspring of Harvard Mice the Product of Unprotected Sex?*, Dayton Intellectual Property Law Association, The University of Dayton, School of Law, Dayton, OH, October 8, 2010.
- S. S. Han, *The BRCA Battle*, Dayton Intellectual Property Law Association, The University of Dayton, School of Law, Dayon, OH, February 12, 2010.
- S. S. Han, *What Nature Giveth, the Law Taketh Away*, Continuing Legal Education, The University of Dayton, School of Law, Dayton, OH, December 21, 2009.
- S. S. Han, *Bickering Over BRCA*, CincyBio 2009, Cincinnati, OH, November 4, 2009.
- S. S. Han, M. Willenbrink, D. M. Lechleiter, *The Importance of Patent Law in Technical Innovation and Entrepreneurship*, Innovation Center, School of Engineering, The University of Dayton, Dayton, OH, March 13, 2009.
- S. S. Han, *Browse-Wrap Agreements in Ohio*, Dayton Intellectual Property Law Association, The University of Dayton, School of Law, Dayton, OH, November 14, 2008.
- K. G. Helmer, S. S. Han, C. H. Sotak, *Correlation Between ^1H Diffusion Coefficient and Oxygen Tension in RIF-1 Tumors*, Abstracts, Oral Presentation, In. Proc. of the 4th International Society for Magnetic Resonance in Medicine (1), 264, New York, NY, 1996.
- S. J. Gemmell, S. S. Han, K. G. Helmer, A. H. Hoffman, P. Grigg, C. H. Sotak, *Anisotropic and Time Dependent Diffusion of Water in Tendon Under Static Loading*, Abstracts, Poster Presentation, 38th Experimental Nuclear Magnetic Resonance Conference, Orlando, FL, 1997.
- T. Q. Duong, J. J. Neil, H. A. Stark, I. Pályka, C. S. Springer, S. S. Han, C. H. Sotak, J. J. H. Ackerman, *Intracerebroventricular Administration of Compartment-Specific NMR Agents*, Abstracts, Poster Presentation, In. Proc. of the 5th International Society for Magnetic Resonance in Medicine (3), 1613, Vancouver, British Columbia, Canada, 1997.

Curriculum Vitae
Sam S. Han, Ph.D., J.D.

2095 Raceway Trail
Beavercreek, OH 45434
(937) 401-0070 • sam.han.esq@gmail.com

- S. S. Han, B. J. Dardzinski, C. H. Sotak, *Reconciling Differences between Tumor Oxygenation Measurements Performed Using ^{19}F NMR Spectroscopy and Imaging of Perfluorocarbon Emulsions*, Abstracts, Oral Presentation, In. Proc. IEEE 23rd Annual Northeast Bioengineering Conference, Durham, NH, 1997.
- F. Li, T. Tatlisumak, S. S. Han, K. Irie, K. F. Liu, J. H. Garcia, C. H. Sotak, M. Fisher, *Apparent Diffusion Coefficient Maps and Histological Analysis of Varying Time Periods of Temporary Focal Brain Ischemia in the Rat*, Abstracts, Poster Presentation, 23rd International Joint Conference on Stroke and Cerebral Circulation, Orlando, FL, 1998.
- S. S. Han, G. Vetek, M. D. Silva, C. S. Springer, C. H. Sotak, *Deconvolution of Restriction Effects on Compartmental Diffusion Using Combined Relaxography and Diffusion Measurements*, Abstracts, Poster Presentation, 39th Experimental Nuclear Magnetic Resonance Conference, Pacific Grove, CA, 1998.
- M. D. Silva, S. S. Han, C. H. Sotak, *Effects of Signal-to-Noise and Parametric Limitations on Fitting Biexponential Magnetic Resonance (MR) Inversion-Recovery Curves Using a Constrained Nonlinear Least Squares Algorithm*, Abstracts, Oral Presentation, IEEE 24th Annual Northeast Bioengineering Conference, Hershey, PA, 1998.
- S. S. Han, G. Vetek, C. S. Springer, C. H. Sotak, *Apparent Diffusion Coefficient of Intra- and Extracellular Water in Yeast Suspension Measured by Combined Diffusion and Relaxography*, Abstracts, Oral Presentation, In. Proc. of the 6th International Society for Magnetic Resonance in Medicine (1), 535, Sydney, Australia, 1998.
- S. S. Han, P. Grigg, K. G. Helmer, A. H. Hoffman, C. H. Sotak, *Characterization of Water ADC Behavior Under Tensile Loading and Recovery in Rabbit Achilles Tendon Using NMR*, Abstracts, Oral Presentation, International Society for Magnetic Resonance in Medicine: Workshop on Magnetic Resonance of Connective Tissues and Biomaterials, Philadelphia, PA, 1998.
- K. G. Helmer, S. S. Han, M. Meiler, C. H. Sotak, *The Use of pO_2 and Diffusion Measurements for Characterizing Treatment Response in RIF-1 Tumors*, Syllabus, Oral Presentation, Workshop on Magnetic Resonance in Experimental and Clinical Cancer Research, St. Louis, Missouri, 1998.

LECTURES AND SPEAKING ENGAGEMENTS

- S. S. Han, *Fundamental Concepts for Applying Intellectual Property*, Project Management and Innovation (MEE 433 / ECE 433), The University of Dayton, Kettering Laboratories, Dayton, OH, March 29, 2012.
- S. S. Han, *Fundamental Concepts for Applying Intellectual Property*, Project Management and Innovation (MEE 433 / ECE 433), The University of Dayton, Kettering Laboratories, Dayton, OH, March 27, 2012.
- S. S. Han, *Survey of Intellectual Property Concepts*, Invited Lecture (CEE 300), The University of Dayton, Chudd Auditorium, Dayton, OH, November 3, 2011.
- S. S. Han, *Business Considerations for Intellectual Property*, Project Management and Innovation (MEE 433 / ECE 433), The University of Dayton, Keller Hall, Dayton, OH, November 1, 2011.
- S. S. Han, *The Power of Intellectual Property when Properly Applied in Business*, Invited Lecture (CEE 300), The University of Dayton, Chudd Auditorium, Dayton, OH, April 28, 2011.

Curriculum Vitae

Sam S. Han, Ph.D., J.D.

2095 Raceway Trail
Beavercreek, OH 45434
(937) 401-0070 • sam.han.esq@gmail.com

- S. S. Han, *Business Drives Everything: Using Intellectual Property as Business Capital*, Project Management and Innovation (MEE 433 / ECE 433), The University of Dayton, Keller Hall, Dayton, OH, March 17, 2011.
- S. S. Han, *Intelligent Intellectual Property*, Project Management and Innovation (MEE 433 / ECE 433), The University of Dayton, Keller Hall, Dayton, OH, September 30, 2010.
- S. S. Han, *Social Justice and the Greater Call*, Law and Leadership Institute, The University of Dayton, Dayton, OH, July 14, 2010.
- S. S. Han, *Product Development, Experiences, and Intellectual Property*, Sears Recital Hall, The University of Dayton, Dayton, OH, February 18, 2010.
- S. S. Han, *Using Intellectual Property as a Business Asset*, Project Management and Innovation (MEE 433 / ECE 433), The University of Dayton, Department of Mechanical Engineering, Dayton, OH, October 23, 2009.
- S. S. Han, *Business Considerations in Legal Matters*, Project Management and Innovation (MEE 433 / ECE 433), The University of Dayton, Department of Mechanical Engineering, Dayton, OH, November 21, 2008.
- S. S. Han, *The Meaning of Christianity*, Christian Legal Society, The University of Dayton, School of Law, Dayton, OH, November 20, 2008.
- S. S. Han, M. A. Morra, *General Discussion on Patent Evaluation*, OFS Fitel, LLC, Norcross, GA, March 8, 2007.
- S. S. Han, *Anatomy of Patent Litigation*, OFS Laboratories, LLC, Somerset, NJ, September 13, 2006.
- S. S. Han, *Becoming Ambassadors*, Asian Christian Fellowship, Georgia Institute of Technology, Atlanta, GA, January 18, 2006.
- S. S. Han, *God's Plan for Outreach and Ministry*, Asian Christian Fellowship, Georgia Institute of Technology, Atlanta, GA, September 28, 2005.
- S. S. Han, *Constitutional Bases for Intellectual Property*, Handong International Law School (HILS), Pohang Korea, March 11, 2005.
- S. S. Han, *The Biblical Plan for Being Witnesses*, Bulldog Christian Fellowship, The University of Georgia, Athens, GA, February 24, 2005.
- S. S. Han, *The Foolishness of the Cross and the Wisdom of God*, Asian Christian Fellowship, Georgia Institute of Technology, Atlanta, GA, February 23, 2005.
- S. S. Han, D. R. Risley, *Patents and Academia*, Office of Technology Transfer, Pohang University of Science and Technology (POSTECH), Pohang, Korea, November 19, 2004.
- S. S. Han, D. R. Risley, *Academic Pitfalls in Patenting*, Office of Technology Transfer, Seoul National University Industry Foundation (SNU-IF), Seoul, Korea, November 17, 2004.
- S. S. Han, D. R. Risley, *Intellectual Property Primer*, Department of Industrial Engineering, Hanyang University, Seoul, Korea, November 16, 2004.
- S. S. Han, *Contrasting the Biblical Jesus with the Cultural Jesus*, Asian Christian Fellowship, Georgia Institute of Technology, Atlanta, GA, October 6, 2004.
- S. S. Han, S. R. Risley, *Malpractice Overview*, Patent Litigation Group Meeting, TKHR, Atlanta, GA, August 18, 2004.
- S. S. Han, D. R. McClure, M. Nguyen, *Problems with Ownership*, Patent Prosecution Group Meeting, TKHR, Atlanta, GA, August 4, 2004.

Curriculum Vitae
Sam S. Han, Ph.D., J.D.
2095 Raceway Trail
Beavercreek, OH 45434
(937) 401-0070 • sam.han.esq@gmail.com

- S. S. Han, T. Deveau, *Prosecution War Stories*, Patent Prosecution Group Meeting, TKHR, Atlanta, GA, May 26, 2004
- S. S. Han, D. R. Risley, *Interplay of Patents and Academic Research*, Office of Technology Transfer, Seoul National University Industry Foundation (SNU-IF), Seoul, Korea, April 30, 2004
- S. S. Han, D. R. Risley, *Patents and Academia*, Office of Technology Transfer, Hanyang University, Seoul, Korea, April 27, 2004
- S. S. Han, *Tension Between Intellectual Property and Academia*, Biomedical Engineering Department, Worcester Polytechnic Institute, Worcester, MA, October 20, 2003
- S. S. Han, *The Continuing Need for Jesus Christ*, Asian Christian Fellowship, Georgia Institute of Technology, Atlanta, GA, 2002.
- S. S. Han, *The Inadequacy of Man and the Sufficiency of Jesus Christ*, Asian Christian Fellowship, Georgia Institute of Technology, Atlanta, GA, 2001.

BOOKS AND CHAPTERS

- S. S. Han, *Characterization of Biological Systems* Via *Relaxometric and Diffusimetric NMR*, University Microfilms International - Dissertation Services / Bell & Howell Co., Ann Arbor, Michigan, 1998.

ACADEMIC ADVISING

- The University of Dayton, School of Law, Dayton, OH
  - 2012 Spring
    - Renee Thayer Wilkerson, Directed Reading
    - Erik Reicis, Directed Reading
    - Patrick Lynch, Law Review
    - Stephen Allhoff, Moot Court
    - Benjamin Christoff, Moot Court
  - 2011 Fall
    - Renee Thayer Wilkerson, Law Review
    - Patrick Lynch, Law Review
    - Stephen Allhoff, Moot Court
    - Benjamin Christoff, Moot Court
  - 2011 Spring
    - Renee Thayer Wilkerson, Law Review
    - Sly Ayoubi, Independent Studies
  - 2010 Fall
    - Darin Barber, Independent Studies, Graduate Studies
    - Caitlin Carreiro, Externship
    - John Prince, Independent Studies, Graduate Studies
    - Brian Sullivan, Independent Studies, Graduate Studies
    - Renee Thayer Wilkerson, Law Review
  - 2010 Spring

Curriculum Vitae

Sam S. Han, Ph.D., J.D.

2095 Raceway Trail
Beavercreek, OH 45434
(937) 401-0070 • sam.han.esq@gmail.com

- Jeffrey Brown, Independent Studies.
- Caitlin Carreiro, Independent Studies.
- Nancy Green, Independent Studies.
- Randall Stevenson, Independent Studies.
- Brian Sullivan, Law Review.
- Jason Williams, Independent Studies.
- April Ward, Law Review.
- 2009 Fall
  - M. Fran Sweeney, Externship.
  - Paul Harkleroad, Law Review.
  - Danielle Hammer, Independent Studies.
  - Walter "Chip" Herin, Law Review.
- 2009 Spring
  - Timothy S. Ganow, Independent Studies.
  - Hannah Livingstone, Independent Studies.
  - Thomas Hahn, Law Review.
- 2008 Fall
  - Rebecca Greendyke, Law Review.

FELLOWSHIPS, SCHOLARSHIPS, AND GRANTS

- *GAANN Fellowship* — 1997 - 1998
- *Travel Grant - International Society for Magnetic Resonance in Medicine* — 1998
- *Michigan Competitive Scholarship* — 1988 - 1992

AWARDS AND RECOGNITIONS

- *Foil Boat Competition, 1st Place* — 2017
- *Newspaper Tower Building Competition, 1st Place* — 2015
- *Nominee for Judicial Appointment to Cobb County Superior Court* — 2007
- *Founding Member - Intellectual Property Board, Georgia State University, Atlanta, GA* — 2004
- *Who's Who in Science and Engineering* — 2002 - 2003
- *CALI Excellence for the Future Award - Georgia Practice and Procedure* — 2001
- *CALI Excellence for the Future Award - Constitutional Law: Survey of the First Amendment* — 2001
- *CALI Excellence for the Future Award - Antitrust Law* — 2001
- *Who's Who: American Law Students - 21st Edition* — 2001
- *National Appellate Advocacy Moot Court Competition Semi-Finalist, Southeast Regional Competition* — 2001
- *CALI Excellence for the Future Award - Federal Courts* — 2000
- *Who's Who: American Law Students - 20th Edition* — 2000
- *Saul Lefkowitz Trademarks Moot Court Competition 3rd Place, Southeast Regional*

Curriculum Vitae
Sam S. Han, Ph.D., J.D.
2095 Raceway Trail
Beavercreek, OH 45434
(937) 401-0070 • sam.han.esq@gmail.com

Competition	2000
• IEEE Student Paper Competition 1st Place (23rd Northeast Bioengineering Conference)	1997
• Dean's List (The University of Michigan)	1992
• Renaissance Man Award	1988

ACADEMIC COMMITTEES AND SERVICE

- The University of Dayton, School of Law, Dayton, OH
    - Colloquium Committee, Member (2009-Present).
    - Moot Court Board, Advisor, (2010-Present).
    - LL.M. and M.S.L. Graduate Studies Committee, Member (2009-Present).
    - International Studies Committee, Member (2009-2010).
    - Judge, Walter H. Rice Moot Court Competition, The University of Dayton, School of Law (2008-Present).
    - Judge, Dayton Bar Association Moot Court Competition, The University of Dayton, School of Law (2008-Present).

INVENTIVE ACTIVITY

- Issued Patents
    - U.S. Patent Number 6,764,053, Object Holder, issued on July 20, 2004, by Han.
    - U.S. Patent Number 8,852,012, Storage at Indoor Golf Driving Range, issued on October 7, 2014, by O'Grady and Han.
    - U.S. Patent Number 9,597,575, Storage at Indoor Golf Driving Range, issued on March 21, 2017, by O'Grady and Han.
- Patent Applications
    - U.S. Patent Application Serial Number 11/456,439, Systems and Methods for Remotely Enabling and Disabling Non-Voice-Related Functions on Portable Communication Devices, filed on July 10, 2006, by Han.

MEMBERSHIPS

• American Intellectual Property Law Association	Member Since 2011
• Dayton Intellectual Property Law Association	Member Since 2008
• State Bar of Georgia	Member Since 2001
• Intellectual Property Section of the Georgia Bar	2001 - 2008
• American Bar Association	2001 - 2005, 2008 - 2010
• American Bar Association - Student Member	1998 - 2001
• Christian Legal Society - Georgia State University	1998 - 2001
• International Society for Magnetic Resonance in Medicine	1994 - 1998

Curriculum Vitae

Sam S. Han, Ph.D., J.D.

2095 Raceway Trail
Beavercreek, OH 45434
(937) 401-0070 • sam.han.esq@gmail.com

OFFICES

Member, Board of Directors   Engineering and Science Hall of Fame, Dayton, OH
(2011-2015)
- Non-profit (501(c)(3)) international organization established to honor engineers and scientists who have made a significant contribution to human well-being

President and Chief Executive Officer   Traton News, LLC, Beavercreek, OH
(2011-2014)
- News-reporting organization
- Founding member and organizer

Member, Board of Directors   OURS, Inc., Los Angeles, CA
(2008-2010)
- Religious non-profit (501(c)(3)) organization for ministry to at-risk youth
- Non-managing member of the Board of Directors

Organizer, Member, Officer   ExhibitView Solutions, LLC, Rome, GA
(2009-2011)
Organizer, Member, Officer   ExhibitView Management, LLC, Rome, GA
(2008-2011)
Organizer, Member, Officer   ExhibitView, LLC, Rome, GA
(2006-2008)
- Founding Member and Organizer
- Software development company
- Advise on intellectual property and outsourcing matters

President and Chief Executive Officer   Sam Han, P.C., Marietta, GA
(2006-2009)
- Founding Member and Organizer
- Pro bono legal cases
- Non-legal public service
- Non-intellectual-property-related legal matters

Organizer, Member, Officer   Adverless, LLC, Marietta, GA
(2006-2008)
- Founding Member and Organizer
- Corporate governance and operation of Internet-related advertising, including operation of <http://www.unhappysociety.com>

Organizer, Member, Officer   Renter Network, LLC, Marietta, GA
(2005-2008)
- Founding Member and Organizer

Curriculum Vitae
Sam S. Han, Ph.D., J.D.
2095 Raceway Trail
Beavercreek, OH 45434
(937) 401-0070 • sam.han.esq@gmail.com

- Developing avenues for sales and marketing of rental property through various Internet resources

Organizer, Member, Secretary   *Breakfast Technologies, LLC, Marietta, GA*
(2005-2009)
- Founding Member and Organizer
- Maintain records for corporation
- Corporate think-tank for generating intellectual property
- Intellectual property valuation and licensing

President and Chief Executive Officer   *Samster, Inc., Marietta, GA*
(2005-2008)
- Founding Member and Organizer
- Corporate governance and operation for Exhibit View, LLC, Adverless, LLC, Breakfast Technologies, LLC, and Renter Network, LLC
- Maintain records and finances for corporation

President   *Asian American Law Students Association, Georgia State University, Atlanta, GA*
(1999-2000)

Secretary   *Intellectual Property Law Society, Georgia State University, Atlanta, GA*
(1999-2000)

SPECIAL SKILLS

- **Trilingual**
    - English (Fluent)
    - Korean (Fluent)
    - Spanish (Proficient)

- **Programming Skills**
    - Fortran
    - Pascal
    - Assembler Code (Intel™ 8086)
    - CAD
    - Spice
    - C

- **Systems Administration**
    - Hewlett Packard™ HP-UX
    - Silicon Graphics IRIX™
    - Sun SPARC SUNVIEW

Curriculum Vitae

Sam S. Han, Ph.D., J.D.

2095 Raceway Trail
Beavercreek, OH 45434
(937) 401-0070 • sam.han.esq@gmail.com

EXTRA CURRICULAR ACTIVITIES

- Sunday School Teacher
    - Bethany Presbyterian Church, Atlanta, GA   2003 - 2006
    - Korean United Methodist Church, Atlanta, GA   1998 - 2001
    - St. John's Korean United Methodist Church, Lexington, MA   1995 - 1996
    - Chinese Christian Fellowship Church, Ann Arbor, MI   1991 - 1992
- Bible Study Leader
    - Bethany Presbyterian Church, Atlanta, GA   2003 - 2006
    - Korean United Methodist Church, Atlanta, GA   1998 - 2001
    - Christian Bible Fellowship, WPI, Worcester, MA   1997 - 1998
    - Chinese Christian Fellowship, Ann Arbor, MI   1991 - 1992
    - Korean Bible Church, Ann Arbor, MI   1988 - 1992
- Geometry Tutor for Harrison High School Student, Acworth, GA   2005
- Calculus Tutor for Pope High School Student, Marietta, GA   2004
- Mentorship Program, Georgia State University, College of Law, Atlanta, GA   2003 - 2004
- Youth Minister
    - Korean United Methodist Church, Atlanta, GA   2000 - 2001
- The Honorable Thomas Tang National Moot Court Competition   1999
    - Sponsor (GSU-AALSA)
- BAR/BRI Student Representative - Georgia State University, Atlanta, GA   1998 - 2001
    - Senior Representative   1999 - 2000
- University Lutheran Homeless Shelter Volunteer: Harvard Square, Cambridge, MA
                                                                                 1996 - 1997
- Cell Group Leader: St. John's Korean United Methodist Church, Lexington, MA
                                                                                 1996 - 1997
- Minister of Transportation: St. John's Korean United Methodist Church, Lexington, MA
                                                                                 1994 - 1995
- Short Term Missionary to Korea   1993
- Physics Student Instructor: The University of Michigan, Ann Arbor, MI   1992
- The University of Michigan Men's Glee Club, Ann Arbor, MI   1989 - 1992
- Motorcycle Safety: Advanced Rider's Course, Ypsilanti, MI   1993
- Rowing
- Chess
- Close-up Magic
- Tennis
- Classical and Folk Guitar

REFERENCES AVAILABLE UPON REQUEST

www.ingramcontent.com/pod-product-compliance
Lightning Source LLC
Chambersburg PA
CBHW031926240526
45464CB00023B/1675